The Twenty Years' Crisis, 1919–1939

Reception and Praise for *The Twenty Years' Crisis*

'Apparently overtaken by events in the very days of its first production, Carr's *Twenty Years' Crisis* has never been more pertinent to the discussion of international relations than it is today: in a world beset by the twin extremes which he excoriated, a craven and short-sighted realism on the one hand, and an unanchored and irresponsible idealism on the other, Carr's astute arguments should be central to our analysis of, and response to, the world of the twenty-first century.' — **Fred Halliday**, *Professor of International Relations, London School of Economics*

'*The Twenty Years' Crisis* is one of those books that somehow never goes out of date. It brings into sharp focus a lot of the core questions that anyone grappling with the complexities of International Relations needs to confront, and it sets a standard of clarity and vigour of prose that has few competitors in the contemporary IR literature.' — **Professor Barry Buzan**, *University of Westminster*

'. . . now is the time to relaunch *The Twenty Years' Crisis* as a basis for rethinking the problem of world order in a time of greater complexity and uncertainty. [Carr's] exposure of the power relations underlying doctrines of the harmony of interests is especially pertinent to a serious understanding of the ideology of globalization today, while his careful discussion of the need to balance power and morality warns against the hypocrisy of contemporary great-power crusading.' — **Professor Robert Cox**, *Emeritus Professor, York University, Canada*

'In the 20th century E. H. Carr was one of the most original and interesting thinkers about international relations. Carr's insights into the nature of international affairs warrant attention. Everyone interested in international politics should read this book.' — **Robert Gilpin**, *Eisenhower Professor Emeritus of Public and International Affairs, Princeton University*

'. . . the net influence of the book . . . is mischievous.' — **Norman Angell**

'. . . brilliantly reasoned.' — **R. W. Seton-Watson**

'A brilliant, provocative and unsatisfying book.' — **Martin Wight**

'Carr is the consummate debunker who was debunked by the war itself.' — **Arnold Toynbee**

'Professor Carr has shown the entire inadequacy of Professors Zimmern and Toynbee: who will demonstrate the entire inadequacy of Professor Carr?' — **Richard Crossman**

The Twenty Years' Crisis 1919–1939

An Introduction to the Study
of International Relations

E. H. Carr

Reissued with a New Introduction and additional material by

Michael Cox
Professor of International Politics
University of Wales
Aberystwyth

palgrave

First published 2001 by
PALGRAVE
Houndmills, Basingstoke, Hampshire RG21 6XS and
175 Fifth Avenue, New York, N. Y. 10010
Companies and representatives throughout the world

PALGRAVE is the new global academic imprint of
St. Martin's Press LLC Scholarly and Reference Division and
Palgrave Publishers Ltd (formerly Macmillan Press Ltd).

ISBN 0–333–96375–X hardback
ISBN 0–333–96377–6 paperback

This book is printed on paper suitable for recycling and
made from fully managed and sustained forest sources.

A catalogue record for this book is available
from the British Library.

Library of Congress Cataloging-in-Publication Data
Carr, Edward Hallett, 1892–
   The twenty years' crisis, 1919–1939 : an introduction to the
   study of international relations / E.H. Carr ; with new introduction
   by Michael Cox, [editor].
      p. cm.
   Includes bibliographical references and index.
   ISBN 0–333–96375–X (cloth) — ISBN 0–333–96377–6 (pbk.)
   1. International relations—History. 2. World politics. I. Cox,
   Michael, 1947– II. Title.
   JZ1305 .C37 2001
   327'.09'042—dc21
                                                          2001036149

10   9   8   7   6   5   4   3   2   1
10   09   08   07   06   05   04   03   02   01

Printed and bound in Great Britain by
Antony Rowe Ltd, Chippenham, Wiltshire

Contents

Acknowledgements

This project began to take shape when I was preparing the manuscript for my own volume, *E. H. Carr: a Critical Appraisal*, published by Palgrave in 2000. Therefore, I first need to thank all those who got me interested in Carr and those who then suggested that we reissue one of his great classics with an introduction and additional material that would not only help a new generation of students understand Carr more readily, but also explain how he came to write *The Twenty Years' Crisis* in the first place, and why it exercised the influence it did – and still does. That great stalwart of the British intellectual scene, Bernard Crick, once admitted that of all the books he had ever read – and Bernard had read quite a few – it was Carr's *The Twenty Years' Crisis* that had possibly influenced him most. Many others would no doubt say the same thing, including Tim Farmiloe, who worked for Macmillan for many years. Though Tim was too young to publish *The Twenty Years' Crisis* (that privilege was left to the future Prime Minister Harold Macmillan), he did work closely with Carr much later, and as a result encouraged me to get involved in my first venture into Carrdom. I would also like to thank Tim's successor, Josie Dixon, as well as Alison Howson, who first suggested that we reissue the volume with flags flying and drum beating, and a painting by Lowry on the cover to set the whole thing off.

Robert Machesney, who is the keeper of the archives at the Macmillan HQ in Basingstoke, deserves special mention too, not only for the wonderful lunch he provided on my visit there, but for making available to me the Carr correspondence with his favourite publishing house. The letters are worth reading for all sorts of reasons, not the least of which are the light they throw on Carr's determination as a successful author to negotiate such excellent terms! I would also like to express my gratitude to the librarians in the Special Collection in The University of Birmingham where the Carr Papers are housed, and which have been added to enormously since 2001 with such documents as the Carr diaries, as well as several previously unavailable letters. I would also like to thank Caroline Ellerby for working so efficiently and helping me get all the material ready for the book.

Moving closer to home, my sincerest gratitude must be extended to my good colleague Professor Ian Clark in the Department of International Politics at the University of Wales, Aberystwyth. As the proud owner of one of the few first editions of *The Twenty Years' Crisis* it was Ian who made it possible for me to make a serious comparison of the original with the later editions of the book, thus hopefully clearing up once and for all any misunderstandings regarding some of the more controversial deletions Carr made when working on the book in the summer and autumn of 1945 before its appearance in 1946. Directly and indirectly, many other colleagues have helped too, but I would like to say a very special word of thanks to several in the 'home of the discipline' at 'Aber', including Ken Booth, the first holder of the E. H. Carr Chair in International Politics, Tim Dunne, who is more than just 'the Boswell of the English school', Richard Wynn Jones, who still has his doubts about Carr on the national question, Andrew Linklater, a worthy successor to Carr as Woodrow Wilson Professor, Brian Schmidt for raising important questions about the 'first great debate', Nick Wheeler for his sheer enthusiasm and who is at last getting the recognition he deserves, Mike Williams for sharing his insights about Morgenthau, Steve Smith for his continued friendship and support, and Colin Wight for his immaculate taste in critical realism and French reds. Gratitude must also be extended to Barry Buzan, Bob Cox, Robert Gilpin and Fred Halliday for supporting the book with their warm words of praise. Thanks as well go to Stefano Guzzini, Charles Jones and Peter Wilson for their many insights into Carr. I would also like to mention the three previous Chairs of the British International Studies Association – Chris Brown, Christopher Hill and Richard Little – for being so supportive during my term of editorship of the Association's journal, the *Review of International Studies,* between 1998 and 2001. In many ways my work for the *Review* and my involvement with Carr seem to have been indistinguishable these last four years.

Finally, a very special mention for my wife Fiona and my four children, Annaliese, Ben, Dan and Nell, for always being there.

Michael Cox
Aberystwyth
Wales

June 2001

Introduction

In 1936 Edward Hallett Carr resigned from the Foreign Office where he had worked for the better part of twenty years to take up the position of the Woodrow Wilson Chair in the Department of International Politics at the University College of Wales Aberystwyth. At first sight it was one of the oddest decisions he ever took in an extraordinarily long and very distinguished career. Aberystwyth, after all, was hardly one of the more fashionable universities in the United Kingdom, and West Wales was about as far away as it was possible to be from the centres of political and intellectual power where Carr had hitherto spent nearly all of his life.[1] Carr also discovered (just after the interview) that the President of the College – the redoubtable Lord David Davies, whose support for the League of Nations was only equalled by his benevolence in funding the post – was totally opposed to his appointment on the grounds that Carr was both sceptical about the League[2] and critical of the liberal principles enunciated by another, rather more important president, after whom the

---

[1] See E. L. Ellis, *The University College of Wales Aberystwyth 1872–1972* (Cardiff: University of Wales Press, 1972), esp. pp. 245–60.

[2] In his Inaugural Lecture delivered in Aberystwyth on 14 October 1936, Carr (who had at one time been Assistant Adviser on League Affairs in the Foreign Office) criticized the idea of using economic sanctions to either punish or deter Japanese aggression in Asia. This infuriated his liberal critics who also happened to be close to Davies. One such was Gilbert Murray, who attacked Carr directly in a personal letter for having in his lecture repudiated 'the whole League principle'. How, he continued, was the expansion of Germany, Japan or Italy to be resisted if sanctions were abandoned? Carr replied that unless one was prepared to resort to force, then all talk of sanctions was pointless. Cited in Christopher Thorne, *The Limits of Foreign Policy: the West, the League and the Far Eastern Crisis of 1931–1933* (London: Macmillan, 1972), p. 382. The Carr Lecture was published under the title of 'Public Opinion as a Safeguard of Peace', *International Affairs*, Vol. XV, No. 6, November–December 1936, pp. 1–17. See also his 'The Future of the League', *The Fortnightly*, October 1936 (pp. 385–96) where Carr contrasts the League of the Realists which he associated with the Covenant 'a cautious and comparatively "realistic" document', and the 'League of the Idealists' who hoped to use the League to prevent war by 'turning sovereign states into good, law-abiding members of the international community'.

Chair had originally been named.[3] The initial signs were hardly encouraging. But, as the philosophical Carr explained in a letter sent to his publisher (the future Prime Minister, Harold Macmillan), the position did have its advantages. As he pointed out, it was the 'best' that there was 'in the country' and 'the obligations of residence' were 'extremely small'. It would also give him what he could never have had in the Foreign Office: the freedom 'to write and lecture on foreign affairs', something he had 'long wanted to do'.[4] This, if nothing else, was the critical attraction of what he later referred to rather puckishly as this 'fancy chair' in the people's College by the sea.[5]

The unlikely encounter between Carr the cosmopolitan and Aberystwyth the university turned out, in fact, to be one of the most productive in academic history. Given the space he craved Carr published not just one book while in the employ of the University of Wales, but no less than seven, a remarkable output even by Carr's already remarkably productive standards. The first (Carr's favourite) sank almost without trace, except for that which it left on the anarchist movement.[6] The next – a straightforward student text dealing with the international history of the inter-war period – went on to sell well over 70,000 copies.[7] It also had the rare privilege of being his only book published in Welsh.[8] At least three more, *Conditions of Peace, Nationalism and After* and *The Soviet Impact on the*

---

[3] The controversy surrounding the Carr appointment in Aberystwyth is explored in Brian Porter, 'E. H. Carr – the Aberystwyth Years, 1936–1947', in Michael Cox (ed.), *E. H. Carr: a Critical Appraisal* (Basingstoke: Palgrave, 2000), pp. 36–67.

[4] Letter from E. H. Carr to Harold Macmillan, 17 March 1936. Carr Papers, the University of Birmingham Special Collection (hereafter CPB).

[5] Carr used the term in an interview with Peter Scott, 'Revolution without Passion', *The Times Higher Educational Supplement*, 7 July 1978, p. 7.

[6] *Michael Bakunin* (London: Macmillan, 1937).

[7] *International Relations Since the Peace Treaties* (London: Macmillan, 1937); later reissued under the new title *International Relations Between the Two World Wars* (London: Macmillan, 1947). The success of the book was in large part due to Carr spotting a special gap in the market. In his 17 March 1936 letter to Harold Macmillan, he believed that there might be 'a big demand for such a book ... not only for WEA [Workers' Education Association] purposes but for the sixth forms of schools (particularly girls' schools where they now all study 'current events')'. There are no separate figures to indicate just how many 'girls' schools' actually purchased the volume! (CPB).

[8] In his 17 March 1936 letter to Harold Macmillan, Carr wrote that 'the first result of the appointment [to the Wilson Chair] is that I have been asked whether or not I will write a short text-book on international affairs since 1919 for publication in Welsh by the university press (mainly I gather, for educational purposes). It will only be worth my while to do so if I can also find an English publisher. Would your firm be interested?'. Carr sent in final proofs in January 1937 (CPB).

Western World, were all best-sellers too,[9] especially the first one, which even the normally modest Carr admitted had been 'a spectacular success'.[10] Unfortunately, the same could not be said about his earlier and more detailed study on British foreign policy. Published in late 1939 it was more or less still-born.[11] The same fate did not await his other volume, however, which appeared at about the same time. Conceived in 1937 with an entirely different title to that which finally appeared on the cover (Carr's preferred option until he was persuaded otherwise was *Utopia and Reality*) it was in the end published – after detailed negotiations between Carr and his publisher – under the name by which it ultimately became internationally famous: *The Twenty Years' Crisis: 1919–1939. An Introduction to the Study of International Relations.*[12]

The history of *The Twenty Years' Crisis* is almost as interesting (and perhaps as significant) as the book itself. As his letters and diaries of the time reveal, Carr never intended to write a popular or even an especially accessible study. As he rather nicely put it in 1938, this particular volume was not one that 'people' would 'want to take to the seaside or the mountains with them'.[13] Nor was it ever written with the mass market in mind, but for what Carr liked to regard as that rather nebulous, but all-important body of human beings known as the intelligent reading public. Indeed, in another communication to Macmillan just after the book was published, he worried that with the onset of the war and the call-up, those for whom the book had originally been intended would not be around to read it.[14] He was equally concerned that the new title which he had only agreed with Macmillan a month or so before publication, did not accurately reflect what was in the book itself. A 'fancy title' would be inappropriate, he agreed.[15] On the other hand, the book was not a 'bigger and better history of post-war international relations' (which was implied

---

[9] *Conditions of Peace* (London: Macmillan, 1942). *Nationalism and After* (London: Macmillan, 1945) sold something close to 30,000 copies and was translated into six languages. *The Soviet Impact on the Western World* (London: Macmillan, 1946) sold nearly as well and was translated into five languages. Figures from the Macmillan Publisher Archives (hereafter MPA).

[10] See E. H. Carr, 'Report And Programme Of Wilson Professor', 6 April 1943, p. 53. The University College of Wales Aberystwyth Archives (hereafter UWA).

[11] *Britain: Study of Foreign Policy from the Versailles Peace Treaty to the Outbreak of War* (London: Longmans, Green & Co., 1939).

[12] Carr signed the contract to write *Utopia and Reality* on 25 July 1938. The title *The Twenty Years' Crisis (1919–1939): an Introduction to the Study of International Relations* was only agreed to on 15 August 1939 (MPA).

[13] Letter E. H. Carr to Harold Macmillan, 15 December 1938 (MPA).

[14] Letter E. H. Carr to Harold Macmillan, 12 December 1939 (MPA).

[15] Letter E. H. Carr to Harold Macmillan, 27 July 1939 (MPA).

by *The Twenty Years' Crisis*) but rather 'about international relations in general'[16] and while he conceded (reluctantly) that *Utopia and Reality* might be too 'abstract', he was always anxious that the final alternative did not give a clear enough indication of what was actually in the volume itself. In terms of sales at least he need not have worried. Under its new and more popular heading, it sold extremely well and was widely reviewed, a fact that pleased the cautious but quietly confident Carr no end.[17] The book, he was happy to report to his College back in Wales in 1943, had achieved 'substantial success'.[18] This not only prompted Macmillan to bring out a new second edition in 1946, but also led to its speedy translation into two foreign languages, to be followed by several more in the years which followed.[19] Whether Carr had intended it or not, *The Twenty Years' Crisis* had become an academic best-seller.

What Carr could never have anticipated, of course, was the book's subsequent influence and minor cult-like status within the new academic discipline of international relations, a subject for which he appeared to have very little enthusiasm at all. Indeed, one of the many ironies of *The Twenty Years' Crisis* is that its author not only demonstrated little interest in its subsequent fate, but showed even less in the discipline which now claimed the book as one of its classical texts. Clearly this had something to do with his post-war preoccupation; writing a massive history of early Soviet Russia. Yet, as the record shows, his indifference was not just because he had changed academic trains. He also had grave doubts about the subject itself. As he explained in 1971 to an American admirer, he had 'long thought that International Relations' was a 'rag-bag' into which one was 'entitled to put anything one pleases', and that the various attempts to 'turn it into some sort of organzied self-contained subject' had

---

[16] Letter E. H. Carr to Harold Macmillan, 15 August 1939 (MPA).

[17] Letter E. H. Carr to Harold Macmillan, 2 December 1939 (MPA).

[18] Professor E. H. Carr, 'Report and Programme of the Wilson Professor', 6 April. University of Wales, Aberystwyth Archives (UWA).

[19] It was planned to translate *The Twenty Years' Crisis* into Dutch in 1940 but the war made this impossible. It was first translated into Swedish in 1941 and Spanish in 1942 (Sweden and Spain being neutral countries). It then appeared in Japanese (in 1951 and 1994), German (1972), Portuguese (1981 and 2000), Greek (1999) and Korean (2000). Information from Carr's literary agents, Curtis Brown of London. Total sales of the book in English are difficult to calculate from the files, but my own estimate is somewhere around 50,000. This includes all the reprints of the 1st edition of 1939 (three in 1940 alone), the thirteen printings of the 2nd edition of 1946, and the Papermac version of the book which came out in 1981, 1983 and 1984 (twice). This figure does not include sales since 1984, foreign language sales, or sales from the US edition published by Harper & Row, New York in 1964.

ended in 'fiasco'.[20] A few years later and he was saying much the same to another American, the distinguished American academic Stanley Hoffman. If anything he was more hostile, and even attacked his own role in having inadvertently helped establish the subject. 'Whatever my share in starting this business' (this 'business' being international relations), 'I do not know that I am particularly proud of it'. Anyway, he asked, what is this thing called international relations in the 'English speaking countries' other than the 'study' about how 'to run the world from positions of strength'? In other places, at other times, it might be something else, but within those states which had the influence – as opposed to those that did not – it was little more than a rationalization for the exercise of power by the dominant nations over the weak. There was no 'science of international relations' he went on. The subject so-called was an ideology of control masking as a proper academic discipline.[21]

Carr's hostility to international relations could not, however, alter the course of academic history, and whether he liked it or not, his apparently unloved child, *The Twenty Years' Crisis*, went on to play a quite crucial role in both launching the subject and influencing the way in which generations of students were taught to think about its history. For in spite of what Carr may have intended when he wrote the book, those who came to read it later – and in particular during the dark, dangerous days of the Cold War – came to regard it as a classic 'realist' text which condemned any talk of ethics or morals and saw all international politics as a simple struggle for power between competitive states in an unchanging international system.[22] This one-sided reading of what was a much more nuanced text, subtly but effectively transformed *The Twenty Years' Crisis* from what Carr had originally set out to do – which was to debunk the pretensions of liberalism while providing a way out of the impasse that was the inter-war crisis – and make it into something rather conservative and simple. *The Twenty Years' Crisis* was not exactly turned into its opposite. On the other hand, it was stripped bare of any critical edge. This did not necessarily make it the least understood book in the history of international relations. But it definitely came close. Certainly, it was never read with any great care, particularly by those who seemed more intent on legitimizing a particular way of thinking about the world rather than studying what Carr actually said.

[20] Letter E. H. Carr to Sheila Fitzpatrick, 4 November 1971 (CPB).
[21] Letter E. H. Carr to Stanley Hoffman, 30 September 1977 (CPB).
[22] See for example, Ieuan John, Moorhead Wright and John Garnett, 'International Politics at Aberystwyth, 1919–1969', in Brian Porter (ed.), *The Aberystwyth Papers: International Politics* (London: Oxford University Press, 1972), p. 97.

However, before going on to look at the very peculiar fate that awaited *The Twenty Years' Crisis*, it would be useful first to return to Carr's original decision to leave the relative security of the Foreign Office to take up that position in that distant university department. We do not do so out of a deference to history – though the history itself is not without its own fascination – but to understand why it was that Carr needed the freedom he so obviously craved to speak out on the major issues of the day. Indeed, only by examining the background to that otherwise odd decision (at least one of his colleagues took it for granted that he had been offered a Chair in Oxford[23]) can we really begin to understand why he decided to write *The Twenty Years' Crisis* in the first place. Books after all do not appear out of thin air, and this particularly influential book, which took him the best part of two years to write, can only be understood if we enter into Carr's world, a world that stood in ruins long before he took the critical decision to depart the Foreign Office and write the study which finally established his reputation as a major thinker of the first order.

E. H. Carr: the unmaking of a liberal

Our story begins not with Carr the newly created professor (a term by the way that he heartily disliked) but Carr the committed liberal whose early world was marked by certainty and security – 'security not only in family relations but in a sense of scarcely imaginable today' he later wrote. As he himself confessed, he grew up in an entirely sheltered environment of public school, Cambridge and the Foreign Office; in a place and time that was 'solid and stable', where prices remained the same, where everybody knew their station in life, in which things only changed slowly, and then only for the 'better'. Even the shock of the First World War and the Bolshevik revolution did little to disturb this equilibrium. The war was a tragedy that destroyed the lives of millions, but, as he recalled, few people at the time 'thought of it as the end of an epoch'. Instead they regarded it, or at least he did, as some 'unpredictable natural calamity' that once over and the damage repaired would leave things more or less as they were before. The same seemed to be true of the Russian revolution. Carr was not foolish enough to think that the new regime would collapse at the very first hint of opposition and resistance, the standard view after 1917. But he never felt that this utopian experiment with what he felt were its

---

[23] In one diary entry, Carr notes that 'Simon' [Sir John Simon] rang up to ask if I had been elected to a chair at Oxford'. 6 February 1936 (CPB).

unrealizable goals of creating a classless society,[24] and spreading proletarian international revolution,[25] would ever amount to very much. No doubt for this reason, although there were others, he did not so much fear the Bolsheviks during the first decade of their rule as feel indifferent to them. As he admitted, until the late 1920s and early 1930s he did 'not give much thought' to what was happening in the USSR. It was, to paraphrase a term used about another country at another time, a faraway place about which he knew something including the language but did not care a great deal.[26]

The first seeds of doubt were sown with the Versailles Peace Treaty in 1919, a key historical event which he attended and in which he played a minor role, primarily dealing with the question of minorities in Central Europe.[27] The big question, however, was what to do with Germany. Carr was by no means naïve There were critical issues that had to be addressed. Nevertheless, these could not be dealt with seriously, and over the long term by punishing the whole nation, and then treating it as a permanent pariah. If nothing else, this made little sense economically, a view he shared with John Maynard Keynes from whom he derived most of his thoughts about the disastrous 'economic consequences of the peace'. It was also bound to store up major problems for the future. Moreover, it was unjust, a common opinion amongst a generation of British intellectuals at the time, including Carr himself who battled long and hard within the Foreign Office to get his position accepted. In the end he failed and, as we now know, felt compelled to resign in order to be able to speak out in public. Nor did his views on the subject change much thereafter, in spite of what happened in 1939 and the failure of appeasement as a policy. As a close confidante remarked many years after his death, Carr never had any real sympathy for Hitler. Nonetheless, he always thought Versailles to have been little short of disastrous and a

---

[24] Under the *nom de plume* of John Hallett, Carr wrote in 1933 that the utopian notion of a classless society was the 'Achilles' Heel of Marxism', and that the idea of one was as 'dull as the heaven of the orthodox Victorian'. See his 'Karl Marx: Fifty Years After', *Fortnightly*, March 1933, p. 321.

[25] For Carr's scepticism about 'world revolution' see his later History of Soviet Russia (1950–1978. London, Macmillan, 10 volumes, 1950–78). Carr recognized the influence of the USSR on the West. However, he never thought that revolution in the West was likely or feasible.

[26] On Carr's early views, see his 'Autobiography' in Michael Cox (ed.), *E. H. Carr: Critical Appraisal*, pp. xiii–xxii.

[27] Alan Sharp, 'Britain and the protection of minorities at the Paris peace conference, 1919', in A. C. Hepburn (ed.), *Minorities in History* (New York: St Martin Press, 1979), pp. 173 and 185 n.12.

major factor in both bringing Hitler to power in 1933 and causing the war a few years later.[28]

If the harsh treatment meted out to Germany in 1919 was one of the reasons for Carr's growing disillusionment, the entry of the United States onto the world's diplomatic stage at Versailles was undoubtedly another. Carr was not especially anti-American.[29] However, he was anti-Woodrow Wilson, and it was his critical engagement with this particular President's brand of crusading politics that raised serious doubts in his own mind about both liberalism as a doctrine and Wilsonianism as a guide to international politics. His hostility to Wilson in particular (which David Davies could never quite forgive)[30] derived from two main sources: a disdain for Wilson's high-sounding moral rhetoric, which Carr regarded as little more than an idealistic fig-leaf masking America's ambition of extending its own influence at the expense of others; and a more principled opposition to Wilson's use of the slogan of the right of national self-determination. Carr was no lover of empires as such. On the other hand, he could detect nothing particularly progressive arising from their collapse in Eastern Europe. Furthermore, as he repeatedly pointed out, the proliferation of a host of small European states in the inter-war period could only add to an already highly unstable international situation. No great fan of small states at the best of times, he soon came to believe that one of the additional reasons for the crisis between the wars was the highly irresponsible way in which the liberal ideal of national self-determination had been applied in 1919. As he noted 'national self-determination was a principle' that not only had 'awkward implications for bourgeois democracy' but for 'international concord' as well.[31]

[28] Mary Pomeroy letter to Jonathan Haslam, November 1995 (CPB).

[29] In early 1940 Carr objected to certain additions being proposed in order to update his *International Relations Since the Peace Treaties*. Significantly, he did so in part because these new 'references to the United States' struck him as being 'both inadequate and to put it mildly, unsympathetic in tone'. He continued 'I should particularly dislike it if a book for which I retain, in any event, a large share of the responsibility, contained passages about the United States which would inevitably provoke and irritate any American reader into whom whose hands they might fall'. Letter from E. H. Carr to Harold Macmillan, 12 February 1940 (MAB).

[30] As part of his ongoing campaign against Carr, Lord Davies raised the issue of Carr's critical attitude towards the American President. Over six years after Carr's appointment, he wrote 'I repeat I don't know what his [Carr's] policies are, but as you suggest they are anti-Wilson, [and] it seems curious that the Professor [Carr] should be content to occupy the Wilson Chair ...' See his correspondence with Gilbert Murray, 5 June 1943 (UWA).

[31] Quote from E. H. Carr, *From Napoleon to Stalin and other Essays* (London: Macmillan, 1980), p. 7.

Potentially progressive in its consequences at an earlier period in time, its results could only be destructive and reactionary after the First World War, turning it from a source of potential liberation from oppression in one age, to what Carr termed the 'world's bane' in another.[32]

Carr's faith in liberalism, however, received its greatest blow from the collapse of the world economy after 1929. In this he was by no means unique but representative of what happened to a whole generation. Yet, once detached from his moorings there seemed to be no going back. 'England' and the world along with it 'was adrift', he wrote in 1930, and unless a new 'faith' could be found there was little hope for the future.[33] He found his faith, of course, in social reform and economic planning, as he made clear to an audience at Chatham House a decade later. There could be no return to an economic order that had so obviously failed. What was now needed, he continued, was not some tinkering with the international superstructure (the basic weakness of such schemes as the League of Nations), but a social revolution from below that would attack the underlying source of the current disorder: namely, the unregulated and unplanned character of a liberal economy which presupposed that the hidden hand of the market would invariably lead to the greatest economic good always being achieved for the greatest number. Such an outlook may have made perfect sense in the nineteenth century in a period of expansion: it made none whatsoever in an era of capitalist decline.[34] As he explained in a later memo, private enterprise in the past had not just rested on a number of economic assumptions but a set of 'moral principles' as well. Now all that had changed, leaving the system that had once 'enjoyed general assent' standing indicted before the bar of history.[35] The task, therefore, was clear: to find an alternative, or what Carr termed a 'new society based on new social and economic foundations' whose first objective would be to promote full employment, equality and social justice. Without such a programme of radical change, the world in general, and Britain in particular, would continue to drift.[36]

If the trauma of the Depression disturbed Carr, what unsettled him even more was the contrast between what he now saw happening in the West and what he began to see taking place in the Soviet Union. No great

[32] John Hallett (E. H. Carr), 'Nationalism: the World's Bane', *Fortnightly*, June 1932, pp. 694–702.
[33] See John Hallett, 'England Adrift', *Fortnightly*, September 1930, pp. 354–62.
[34] Professor E. H. Carr, 'What Are We fighting For?' Record of meeting held at Chatham House, 14 August 1940 (CPB).
[35] Untitled draft dated June 1944, 12pp (CPB).
[36] Memorandum from Mr E. H. Carr to Mr Barrington-Ward, 5 August 1940 (CPB).

supporter of the regime in its earlier, more revolutionary days, the coincidence of the capitalist crisis and the apparent successes being enjoyed by the Five Year Plans after 1929 compelled him to rethink his views.[37] This in part was facilitated by his growing love affair with Russian culture more generally.[38] Indeed, it was his work on such nineteenth-century figures as Dostoevsky, Herzen and Bakunin (alienated intellectuals who lived 'outside' of what Carr called 'the charmed circle') that initially led him to think more critically about what he called 'western ideology'.[39] But this was only a precipitating factor. More important by far was what was happening within the USSR itself as it moved from the relative quiet of the 1920s to the hectic drive for modernization that characterized the 1930s. Carr never became an apologist for Stalin or the Soviet Union. On one particularly famous occasion he was even moved to observe that Germany under Hitler 'was almost a free country' by comparison with its totalitarian counterpart to the East. But he always retained what he liked to think of as balance, and consequently felt that in spite of everything it was still important to recognize the advances that were being made at the economic level.[40] This did not make the USSR a socialist society. Neither did it make it a natural ally of the West. But one could not close one's eyes to what was going on, and no amount of theoretical criticism from classical liberal economists like von Mises or Hayek could alter the fact that under conditions of planning things were being successfully undertaken that were not being done in the West – and the West could only ignore all this at its peril.[41]

Finally, as an increasingly unsettled intellectual, Carr was bound to be influenced by new ideas, and as one set of truths collapsed he logically began to look around for another with which to make sense of a world in crisis. Not surprisingly, in the context of the 1930s, this led him to engage more seriously with Marx and Marxism. The outcome was a complex amalgam that managed to combine his own elitist scepticism about

[37] See, for example, John Hallett (E. H. Carr), 'The Poets of Soviet Russia', *The Christian Science Monitor*, 25 April 1929, and 'The Soviets Through Soviet Eyes', *The Spectator*, 14 September 1934.
[38] See E. H. Carr, *Dostoevsky, 1821–1881* (London: Unwin Books, 1931), *The Romantic Exiles* (London: Victor Gollancz, 1933) and *Michael Bakunin* (London: Macmillan, 1937).
[39] 'An Autobiography', pp. xvi–xvii.
[40] Professor E. H. Carr, 'Impressions of a Visit to Russia and Germany'. Lecture given at Chatham House, London, 12 October 1937 (CPB).
[41] For a detailed analysis of Carr's position or positions on the USSR see R. W. Davies, 'Carr's Changing Views of the Soviet Union', in Michael Cox (ed.), *E. H. Carr: a Critical Appraisal*, pp. 91–108.

socialism and doubts about many aspects of Marxism (including the labour theory of value and the notion of class struggle) with something that could loosely be described as being radical and materialistic without ever being properly defined as Marxist itself. It was by no means a straightforward or happy marriage, as his own rather odd biography of Marx demonstrated.[42] Yet even this study demonstrated his respect for Marx as a social scientist and theorist of capitalist crisis.[43] Carr was equally impressed by certain writers who had been influenced by Marx without necessarily being Marxist themselves, especially the sociologist Karl Mannheim and the radical theologian Reinhold Niebuhr.[44] There is no doubt that Carr studied both in depth, and the two together obviously exerted what Carr later agreed was the 'strongest influence' possible 'upon his thinking'.[45] Perhaps for this reason, amongst others, he confessed that although his 1939 study was 'not exactly a Marxist work', it was nonetheless 'strongly impregnated with Marxist ways of thinking, applied to international affairs'.[46]

Constructing a classic: *The Twenty Years' Crisis*

The world that Carr had known therefore was in turmoil and in many ways *The Twenty Years' Crisis* gave expression to this. As someone later remarked, while Carr's classic might have had some important things to say about international relations in general, it was very much marked by the 'diseased situation' during which it was been conceived.[47] Certainly,

[42] E. H. Carr, *Karl Marx: a Study in Fanaticism* (London: J. M. Dent, 1934).
[43] John Hallett (E. H. Carr), 'Karl Marx – Fifty Years Later', *Fortnightly*, March 1933, pp. 311–21.
[44] Two of the books which influenced Carr most were Reinhold Niebuhr, *Moral Man and Immoral Society: a Study in Ethics and Politics* (New York: Charles Scribner's Sons, 1932) and Karl Mannheim, *Ideology and Utopia: an Introduction to the Sociology of Knowledge* (London: Kegan Paul, Trench, Trubner & Co., 1936). In 1949 Carr spoke warmly of Niebuhr's critical insights into the modern condition. See his 'The Moral Foundations for World Order', in E. L. Woodward et al., *Foundations for World Order* (Denver, Colorado: University of Denver Press, 1949), pp. 58–9. In 1953 Carr also wrote a highly sympathetic appreciation of Mannheim. Carr's description of Mannheim could have almost been autobiographical. He wrote that while Mannheim 'was never a Marxist ... the Marxist foundations of his thought went deep'. See his 'Karl Mannheim' in E. H. Carr, *From Napoleon to Stalin*, pp. 177–83. See also Charles Jones, 'Carr, Mannheim and a Post-Positivist Science of International Relations', *Political Studies*, Vol. XLV, 1997, pp. 232–46.
[45] Quote from his letter to Stanley Hoffman, 30 September 1977 (CPB).
[46] 'An Autobiography', pp. xvi–xvii.
[47] Martin Wight, *International Theory: the Three Traditions* (Leicester: Leicester University Press, 1991), p. 267

Carr's analysis of why there was a crisis was never entirely consistent.[48] At times he seemed to lay most of the blame on liberalism itself, at others on the quality of western diplomacy, occasionally on the selfishness of what he called the 'have powers' and their refusal to contemplate any change in the status quo, sometimes on the economic collapse of the 1930s, and by implication on the fundamental contradiction of maintaining any form of equilibrium in an international system composed of nation-states. As such *The Twenty Years' Crisis* combined both sharp analysis and acid condemnation in equal measure, which not surprisingly made Carr more than a few enemies. The volume also represented a very special kind of intervention into the foreign policy debates of the day in which Carr had already been closely involved while in the Foreign Office. Carr, to be fair, never sought to hide his views nor apologize for the fact that he favoured some form of détente with Germany. However, he did not do so for the normal conservative anti-Soviet reasons that favoured Germany as the best bulwark against communism, but because he felt that there was really no serious alternative. As a declining power Britain, he believed, faced one of two imperfect choices in an imperfect world: either to oppose any alteration within the European status quo and thus run the risk of war with a potentially superior enemy at a time when neither the United States nor the USSR could be counted upon,[49] or to accept the necessity and inevitability of some measured change, and having come to that conclusion, seriously engage with Germany in order to bring it about by peaceful means. The tragedy was that it did neither. The result was disastrous. Unwilling or unable to fight a war or even prepare for one, Britain was compelled to retreat and by so doing fed the dangerous illusion in Germany that it could continue to expand with impunity. By then failing to accept that the status quo was in need of adjustment, and that Germany did in fact have legitimate grievances, British policy-makers only managed to annoy the Germans and exacerbate rather than improve relations between the two countries.[50]

Carr's immediate reason for writing *The Twenty Years' Crisis*, therefore, was a most practical one. But he also saw himself engaged in something

[48] See Whittle Johnson, 'Carr's Theory of International Relations', *The Journal of Politics*, Vol. 29, 1967, pp. 861–84. Johnson is right to suggest that *The Twenty Years' Crisis* does not present a unified view of this crisis. He is wrong to say that Carr therefore has 'two theories of international relations'.

[49] As early as October 1937, Carr was anticipating a possible rapprochement between the Soviet Union and Nazi Germany. See his 'Impressions of a Visit to Russia and Germany', pp. 27–8

[50] Carr discusses the British security dilemma in his *Britain: a Study of Foreign Policy*.

far more important than a mere defence of appeasement (the underlying purpose of the chapter on 'peaceful change') or in attacking an increasingly redundant liberalism and certain liberal writers whose only function it seemed was to provide him with a steady supply of ready-made targets which he could then shoot down at regular intervals. His aim, essentially, was to write a more philosophical work that reflected on the nature of knowledge itself and the relationship between different types of knowledge on the one hand, and history and the production of ideas on the other. This is why his original title for the book – *Utopia and Reality* – was so significant. He was adamant. He was not attempting to publish a more detailed or higher level version of his already popular international history published in 1937. Nor was he interested in writing something on British foreign policy and its failures. What he was seeking to do was something else: to explore the interplay between the world as understood by those whom he defined in a fairly loose sense as 'utopians' – a miscellaneous group whose only uniting feature was a desire to project their own thoughts about the world on to the world itself – and those whom he called 'realists', whose chief defining characteristic was an empirical approach to the world as it was. According to Carr, utopianism and realism not only represented 'two methods of approach' but also determined 'opposite attitudes towards every political problem'. Thus the utopian, he asserted, was voluntarist, the realist determinist. The utopian made political theory a norm to which political practice ought to conform, the realist regarded political theory as a sort of codification of political practice. Utopianism attracted intellectuals, realism the bureaucrat. The utopian believed in reason, the realist in force. Utopianism was a trait normally (but not always) associated with the radical, realism with the conservative. Finally, and most fundamental of all, the antithesis of utopia and reality was rooted in a different conception of the relationship of politics and ethics. Hence, the utopian would set up an ethical standard which purported to be independent of politics and sought to make politics conform to it. The realist insisted that morality was a function of politics. In fact, whereas the utopian assumed that there was a general morality and a universal standard by which we could judge what was either right and good, the realist believed that there were no absolute standards save that of fact, and that morality therefore could only be relative and not universal. Viewed from this perspective the search for ethical standards outside of time and place was itself utopian and was thus doomed to frustration.

Why Carr began a book on international relations in this particular way is not at first sight obvious. Nor is it clear initially why he spent the time he did discussing something of which he was so critical as utopianism. Yet as the study proceeds along its extraordinarily erudite way,[51] the reason for opening with a rather abstract discussion about one particular way of viewing the world starts to make more sense. As Carr suggests in the early stage of *The Twenty Years' Crisis* (and as he made clear when he was writing the book), this particular volume was going to be a discussion of what he called 'fundamental trends in international politics' and not some 'topical' work simply dealing with the current situation.[52] It was thus necessary to start not with the immediate crisis but with what he hoped would be a more profound assessment of its underlying causes, one central part of which in his view was the failure of both intellectuals and policy-makers alike to understand the world around them. The question was, why? It was at this point that he turned to the issue of utopianism and the unfortunate influence which he felt it had exerted on the new science of international relations in the period immediately after the First World War.

According to Carr, international relations in its early incarnation bore the birthmark associated with its ethical origins. This was quite under-standable. After all, the war had been a human catastrophe and it was not surprising that those involved in international politics hoped to avoid another war by promoting peace and the institutions associated with peace, especially that most moral of institutions with which the subject in its initial manifestation was to become so inextricably entangled, the League of Nations. But the problem with this approach was obvious: it substituted prescription for analysis while doing little to help explain the causes of peace, or why certain states, and not others, acted in a manner deemed to be aggressive. It also led to an almost complete disregard for power and the role played by power in international affairs. Assuming as they did that there was an international community, and that reason rather than material and military capabilities governed the actions of states, the utopians floundered around in a world of 'make believe' made up of firm resolutions passed by the League of Nations against war but very little else. This made very little difference in the 1920s, when the world was still relatively orderly. It made a great deal, however, after the critical year of 1931 when Japan invaded Manchuria, followed two years later by the rise of Hitler. Now, tired old legalistic slogans about peace and obedience to international law sounded like so much hot air as the revisionist states began to challenge the status quo favoured, not surprisingly, by the

---

[51] 'For by contrast with Carr's other works, *The Twenty Years' Crisis* is drenched in erudition'. Charles Jones, *E. H. Carr and International Relations: Duty to Lie* (Cambridge: Cambridge University Press, 1998), p. 46.
[52] Letter from E. H. Carr to Harold Macmillan, 31 May 1939 (MAB).

remaining democracies in Europe and the European colonial powers in Asia.[53]

Carr's opening move therefore made it possible for him to develop what he hoped would be a theoretically more persuasive analysis of the crisis by showing the extent to which it was being determined by the changing distribution of power in the international system. What it also permitted him to do was draw what seemed on the surface to be a very sharp distinction between one approach, which he saw as being flawed, and another which he did not. But this distinction turned out to be more apparent than real.[54] Hence, while Carr at the beginning painted a rather black and white picture of two modes of thinking, placing himself not surprisingly in the intellectually superior realist category, as his discussion moved forward this early picture of hostile camps composed of aspirational utopians on the one side and cynical realists on the other, began to give way to something rather more subtle and blurred. In fact, as the book more generally wended its complicated way from early thoughts about the new science of international relations to complex assessments about the role of power and morality, all that originally seemed fixed and well-defined began to assume a rather different shape and form.

The first complication was introduced by Carr himself. Having set up his ideal types he then sought to muddy the waters by implying that these twin poles were not opposites at all, but merely two sides of the same intellectual coin. Indeed, as his analysis unfolds, it becomes increasingly clear that Carr neither favours realism in its pure form nor even thinks that such a thing is possible or even desirable.[55] Hence, while he accepts that realism is a necessary corrective to what he calls the exuberance of utopianism, realism without purpose, he argues, is nothing more than 'the thought of old age'. Then, as if to make the point clearer than the truth, he devotes a whole chapter to exploring the limits of realism, and while we are left in little doubt where Carr's intellectual sympathies still

---

[53] The issue of power and security, and how to employ force in the cause of peace, was one that increasingly preoccupied the thoughts of a number of Carr's 'utopians', including Lord David Davies. Throughout the 1930s others like Robert Cecil and Gilbert Murray also began to grapple with the same issue. See Martin Ceadel, *Semi-Detached Idealists: the British Peace Movement and International Relations, 1854–1945* (Oxford: Oxford University Press, 2000), pp. 276–7.
[54] See the essays in Peter Wilson and David Long (eds), *Thinkers of the Twenty Years' Crisis: Inter-War Idealism Reassessed* (Oxford: Clarendon Press, 1995).
[55] 'It is thus rather curious that Carr remains celebrated for his staunch realist critique of idealism'. Stefano Guzzini, *Realism in International Relations and International Political Economy: the Continuing Story of a Death Foretold* (London: Routledge, 1999), p. 22.

lie, he is quite clear in his own mind that realism alone cannot provide a complete basis for effective political thinking. In the end, it is only by drawing upon the visionary aspects of utopianism and the analytical tough-mindedness of realism, that one will be able to develop what Carr refers to as 'sound political thought and a sound political life'.

But this in turn raises a much larger question: what precisely did Carr actually understand by realism? The answer is by no means obvious. For having at one point implied that 'realism' was more likely to appeal to the conservative rather than the radical, he then cited a whole list of modern realists who were not conservative at all. This included a number of German thinkers, including Hegel – from whom he drew much intellectual inspiration – Marx and Engels, whom he cited on several occasions, and Lenin, whom he quoted favourably at a number of points. This was hardly the most obvious combination of well-known conservative thinkers. But then Carr did not see modern realism as being an especially conservative methodology. On the contrary, it was in his view a 'progressive', even subversive outlook that understood all phenomena not as the expression of some general essence standing outside of time but as something historically specific. As Carr observed in a famous section of the book entitled the 'relativity of thought', the supreme virtue of realism was that it allowed the realist to demonstrate that the intellectual theories and ethical principles of one age were all historically conditioned. It also allowed the contemporary realist like Carr to expose the fact – which was critical in terms of his attack on the utopians – that the various ideas of any individual or particular group like the utopians (such as peace and law and order) were not just neutral notions, but rather the expressions of specific interests; and the most urgent task of realism, he went on, was 'to bring down the whole cardboard structure of utopian thought by exposing the hollowness of the material' out of which it was built. Only when this had been accomplished, could we begin the task of finding a way out of the current impasse in which the world now found itself.

If Carr had a particular understanding of realism he had an almost unique notion of what constituted utopianism. For in *The Twenty Years' Crisis* he did not employ it to refer to those who contemplated the possibility of an alternative society somewhere, at some point, in the future. Instead, the utopians of whom he spoke had no vision about the future at all, but rather hung on to an outmoded set of liberal ideals whose principal claim, according to Carr, was that there existed a basic 'harmony of interests' which tried to turn competitive states into members of the same international community. This, he believed, was a simple denial of the basic facts of life in the real world. It was also politically disingenuous.

After all, if there was such a 'harmony' then clearly those who questioned it were by definition trouble-makers, while those who defended it were by implication the peace-makers. But then what was this ideology of 'harmony' other than an ideology justifying a status quo, founded upon a vast inequality of power between the different states? Furthermore, as he pointed out, the utopians so-called did not protest too loudly when the victorious powers imposed their own very harsh peace on the defeated in 1919. However, they made an enormous amount of noise when the 'have-not' powers began to strike back a few years later. This not only smacked of gross hypocrisy, it was also very dangerous for it led to an uncritical support of a situation that was neither defensible not legitimate. Some means would have to be found therefore to permit significant reform to the international system. If not, there could only be one outcome – and that outcome would be war.

The issue of reform then led Carr to look at the problem of 'peaceful change'. He opened his analysis with a discussion of law, treaties and the judicial settlement of international disputes. Given his own outlook, it was hardly surprising that he dealt with each in a way that was least likely to please those who talked rather too easily in his view about the rule of law, the sanctity of treaties, and about the imperative need to resolve international disputes through the various bodies set up under the terms of the League of Nations. Not only did fine words and worthy bodies do little to resolve the outstanding issues; they actually made them worse by fostering the illusion that deep problems could be solved through an appeal to something which did not exist – that is, an international community composed of a society of states who all shared the same goals. This was a chimera and would do nothing to address the underlying causes of conflict which were rooted in the unequal distribution of power. Nor in some ways should one try to arbitrate, for the purpose of arbitration through law was not to change things, but to keep them the way they were. Thus what was required was not some recourse to supranational bodies or courts which had neither the legitimacy nor the power to enforce their decisions, but an honest recognition of the supremacy of politics. Indeed, only by recognizing that all conflicts were ultimately about politics and about the fundamental problem of who gets what, where and when, could one even begin to tackle the problems of the day.

From this assumption Carr drew two important conclusions. The first was that change was both natural and necessary: the task therefore was to ensure that it occured peacefully rather than violently. The second was an acceptance that all change in the end was a function of power rather than morality, and that it was pointless and dangerous standing against the tide of history if the conditions of power determined it. This is why it was essential to concede (up to a point) to the rising power that was Germany. Yet even short-term concessions would not begin to answer the biggest question of all: namely, what sort of international order was required to achieve genuine security over the longer term? Here Carr looked back in order to look forward, and the picture he painted of the future went much further than that ever contemplated by his more utopian enemies.[56] Two reforms of the most basic character were required, he believed – one that would have to tackle the problem of the nation-state, and the other the structure of the international economy. The two issues were intimately connected. Hence if the new world was to be a more peaceful place then it was essential to develop an entirely different approach to economics, and if it was to be more prosperous then it could only become so if people were prepared to rethink the notion of sovereignty, and perhaps, in time, the idea of the nation itself. One thing was clear, however. Liberal schemes with their conceptions of a world federation or blueprints of a more perfect League of Nations would not suffice. These neither attacked the deeper sources of the disorder nor provided any serious solution to the dilemmas facing humanity on the eve of the Second World War. When, and only when policy-makers looked beyond these elegant superstructural answers and began digging in the foundations of the international system, could there be serious historical progress. But, as Carr concluded, this remedy, though utopian at one level, was profoundly realistic at another, for instead of being imposed on history, it grew out of existing historical trends. This is what made his vision of a 'new international order' as opposed to that which had been proposed in 1919, a real possibility.

Liberal critics: Carr's response

As Carr anticipated, the appearance of *The Twenty Years' Crisis* set off a lively and, in some cases, an increasingly bitter discussion amongst those against whom the book had been aimed. It might have helped his case if he had completed his project just a little earlier, but by the time it appeared Britain was at war, and part of his argument about the need to come to terms with Germany had been rendered obsolete by the march of events. Carr, typically, accepted the judgement passed by history – the

---

[56] For a later elaboration of Carr's vision for the post-war world see 'Memorandum From Mr Carr To Mr Barrington–Ward', 5 August 1940 (CPB)

'world's only reliable court' he once observed, quoting Hegel – and within weeks had thrown himself into the war effort, no doubt tinged by some regret that the peaceful change he had argued for had finally given way to a conflict he had hoped might be avoided. Whether he was quite so shocked by what happened as he later seemed to imply was not so clear at the time.[57] In many respects, he had already come to the unpalatable conclusion that there was no possibility of accommodating Germany. He made the point abundantly clear in fact in his other book published at about the same time. Begun in late 1938 and completed by the end of June 1939, this hard-hitting study pulled very few punches in its defence of appeasement. Carr even spared more than a few positive words for the much criticized Neville Chamberlain whose policy of 'realism' he believed was preferable to the wavering course which had been pursued by his predecessors. Yet Carr was never a blind enthusiast, and by March and April 1939 he realized that the game was up. War was now inevitable, more or less, and the nation had to prepare for the long struggle that lay ahead fought on the traditional British principle of preventing what even the normally diplomatic Carr defined as the 'brutal domination' of the continent by a 'single overwhelming Power'.[58] The pity, according to Carr, was that having accepted the logic of their own situation by early 1939, British policy-makers did not then do more to secure the one alliance that might have made the war less likely or at least less costly: that which he was now advocating with Stalin's Russia.[59]

These events, and Carr's speedy return to London to work initially for the Ministry of Information, and then as Assistant Editor for *The Times*, did not prevent him from keeping an eye on the fate of *The Twenty Years' Crisis*. Too much effort had gone into the book for it now to be ignored. Certainly, it did not lack for admirers. The reviewer in the *Times Literary Supplement*, for instance, accepted that even though it was a highly 'provocative' volume, there was no doubting that it was a 'profound' work whose author should be praised for his 'fresh and fearless thinking'. Writing under an assumed name, the up and coming Labour politician, Richard Crossman, was equally impressed. Carr's analysis of the utopians, in particular, was a 'brilliant success' and exposed the 'enervating impact'

[57] In 1980 he recalled that 'the war came as shock which numbed the thinking process' cited in 'An Autobiography', Michael Cox (ed.), *E. H. Carr*, p. xix.
[58] See his *Britain: a Study of Foreign Policy From the Versailles Treaty to the Outbreak of War* (London: Longmans, Green & Co. 1939).
[59] Carr later wrote: 'I remember seeing the guarantee of 1 April 1939, to Poland as the final recipe for disaster. We could not possibly implement it except in alliance with Russia and it was given in such a way to as to preclude any agreement'. See 'An Autobiography', Michael Cox (ed.), *E. H. Carr*, p. xix.

which their ideas had had upon the victorious Powers who, instead of using their power in defence of the status quo or for the accomplishment of peaceful change, had engaged in what Crossman called 'unilateral psychological disarmament'. The American historian, William Maddox, was even more fulsome in his praise. *The Twenty Years' Crisis*, he opined, was nothing less than a 'monument to the human power of sane and detached analysis' that provided a discussion of the 'collapse of the international system' that was entirely 'devoid of national bias'; there was little question in his mind at least that 'Professor Carr' had 'produced one of the most significant contributions to the systematic study of the theory of international politics' that he had 'seen in years'. Another American reviewer agreed and conceded that this 'professor of international politics' from Aberystwyth in Wales had written a 'very forthright' study. Americans would be especially interested, he thought, in Carr's tough assessment of Woodrow Wilson, and while careful not to endorse Carr's critique completely, he hinted strongly that the discerning would find his 'withering criticism' of the crusading President particularly 'stimulating'. The noted sociologist and future theorist of revolutions, Crane Brinton, was equally impressed, though he went on to warn the many US readers of the popular *Saturday Review* that 'when British academic liberals' (by whom he meant Carr) 'begin to go hardboiled, we are entitled to think that something is about to happen'.[60]

Carr was more than a little gratified by these initial comments and like any normal author hoped that these favourable reviews would have a positive impact on sales.[61] Not that he had very much time himself to reflect on these things. The war had changed everything and the principal task now he felt was not to look back on a crisis that had finally been resolved, but to work out the most effective ways possible of integrating British war aims with its plans for the peace. On one issue, however, he was insistent: that it was necessary to learn the lessons of the past and abandon any faith in those various grand plans that had come to nothing after the First World War. It was also critical that the war should not just

[60] These reviews can be found in *Times Literary Supplement*, 11 November 1939, p. 65; Richard Coventry (Crossman), 'The Illusions of Power', *New Statesman*, XVIII, November 25, 1939; *Foreign Affairs*, Vol. 18, No. 3, April 1940, p. 564; and Crane Brinton, 'Power and Morality in Foreign Policies', *Saturday Review*, 17 February 1940, p. 19. The review of *The Twenty Years' Crisis* by William P. Maddox can be found in *American Political Science Review*, Vol. 34, No. 3, 1940, pp. 587–8.
[61] See his letter to Macmillan where he criticizes his publishers for not maintaining stocks of *The Twenty Years' Crisis* in the London bookshops! 2 December 1939 (MAB).

be fought on a moral platform alone with the simple 'negative' aim of destroying 'Hitlerism'. This would neither mobilize the people nor provide them with a vision for the future. If the war was to be fought at all – and fought successfully – it would have to be conducted in the form of what Carr called a 'social war' against privilege and inequality. Politically this was critical. It was also essential if the Allies wanted to 'stir up discontent in the conquered countries against the Nazi yoke'; and they were 'much more likely to' achieve this 'by holding out the prospect of bread and butter than by promises of freedom and self-determination'.[62]

While Carr reflected on the future, his many critics contemplated their next move. One such was the Director of Studies at Chatham House, Arnold Toynbee, one of Carr's main targets.[63] Toynbee fully accepted the power of Carr's polemic against good liberals like himself. But, he wondered, had not this 'consummate debunker' been thoroughly 'debunked by the war itself'? Moreover, no amount of clever intellectual footwork from the admittedly brilliant author of this 'very important contribution to the study of recent international affairs' could disguise the fact that there was a profound 'moral vacuum' at the heart of *The Twenty Years' Crisis*. Criticism was all very well, but unless it led 'to a clearer view of what' was 'morally right and morally wrong' it would in the end lead nowhere. This particular theme was taken up and developed by yet another of Carr's known political enemies, Alfred Zimmern, the first holder of the Woodrow Wilson Chair and now Montagu Burton Professor of International Relations at Oxford. He too was disturbed by Carr's lack of moral compass. Carr was a 'thorough-going relativist', he concluded, who had no way of knowing, or even judging, what was good or bad, right or wrong or plain indifferent. And if there were no such objective standards, he asked, then how could anybody, including Carr, arrive at any reasoned judgement about the world?

One of Carr's many other targets, the well-known liberal theorist, Norman Angell, was even more outraged. *The Twenty Years' Crisis*, he felt, was as 'completely mischievous a piece of sophisticated moral nihilism' as he had ever read. Angell agreed that Carr had forced his opponents (including himself) 'to take stock'. But what was one supposed to make of a book which assumed that law, order, and peace were not general interests but simply the particular interests of the rich and the powerful? Ultimately, all that Carr had done, and presumably done quite

---

[62] Quotes from his memorandum to Barrington-Ward, 5 August 1940 (MAB).
[63] Carr had earlier attacked Arnold Toynbee's views on world politics in *International Affairs*, Vol. XVI, March 1937, pp. 281–3.

consciously, was to provide 'aid and comfort in about equal degree' to the followers of Karl Marx and Adolf Hitler. The Fabian critic, Leonard Woolf, was equally incensed, but expressed himself in a slightly less intemperate way. However, instead of attacking Carr where others had already attacked him before (on the twin issues of relativism and morality) he decided to interrogate what he regarded as Carr's misuse of the concept of utopianism. In Woolf's view Carr never employed the term with any great degree of consistency. Nor did he once think that the same criticism that he had levelled against others might also be levelled against him. Yet if it was legitimate to label as 'utopian' all those who had supported the League, then was it not equally reasonable to attack Chamberlain for being a utopian as well? After all, his policy of appeasement was no more successful at stopping the slide towards war than the League but Carr never lambasted him with the same degree of fervour that he had reserved for his liberal opponents. This not only smacked of hypocrisy but pointed to a basic flaw in Carr's logic, one that he could never admit to of course, because in the end he was guilty of the same inconsistency himself.[64]

Faced with such a barrage of criticism Carr did not respond directly. To that extent there was no great debate between Carr the 'realist' and his 'idealist' critics.[65] What he did do, was to write yet another book, and while it did not represent a 'reply' in any formal sense, it was obvious that part of the reason for publishing *Conditions of Peace* in 1942 was because he had been stung by the various attacks launched by critics. Carr indeed admitted as much a few years later. As he put it, he began to feel 'a little ashamed of the harsh realism of *The 20 Years' Crisis*', and so decided 'to write' what he called 'the highly utopian *Conditions of Peace*'. Admittedly, it was not one of Carr's more profound studies (he wrote it just under a year). On the other hand, it was by no means as 'feeble' as he claimed.[66] Nor was the intellectual distance between it and his previous book as great as one of his reviewers seemed

---

[64] For a more detailed discussion of Carr's early critics see Peter Wilson, 'The Myth of the "First Great Debate"', in Tim Dunne, Michael Cox and Ken Booth (eds), *The Eighty Years' Crisis: International Relations 1919–1999* (Cambridge: Cambridge University Press, 1998), pp. 1–6, and his 'Carr and his Early Critics: Responses to *The Twenty Years' Crisis*, 1939–1946', in Michael Cox (ed.), *E. H. Carr: a Critical Appraisal*, pp. 165–97.

[65] See Brian Schmidt, *The Political Discourse of Anarchy: a Disciplinary History of International Relations* (New York: State University of New York Press, 1998).

[66] See 'An Autobiography', in Michael Cox (ed.), *E. H. Carr: a Critical Appraisal*, p. xix.

to suggest.[67] It addressed the same broad issue: the crisis of the old order. It was equally negative in its evaluation of liberalism. It was vituperative against Wilson. And it did at least try to explain why the 'dissatisfied' powers like Germany had been so dissatisfied and why the satisfied powers after 1919 like Britain had been the authors of so much of their own undoing in the 1930s. It also happened to cite *The Twenty Years' Crisis* on at least one occasion,[68] and was attacked in an equally vituperative way by critics, though this time not for its relativism so much as its indifference to the fate of small nations and overly friendly attitude towards the Soviet Union. One politician was even led to the rather apocalyptic conclusion that with the publication of *Conditions of Peace* Carr had become 'an active danger to the country'.[69]

Carr's other response to *The Twenty Year's Crisis* was less to his various critics and more to what he saw as an unresolved tension in the book. As he later noted, he had in 1939 hovered between two positions which he had never quite resolved in his own mind: one which continued to insist that the nation-state would still remain the principal unit of the international system, and another which hinted (no more) that there could be no sense of order within that system so long as the nation-state remained in being.[70] A series of events in the meantime, including the experience of wartime planning and further reflection on the future of Europe, had resolved that contradiction. By 1945 he had thus arrived at the conclusion that not only was the small nation-state an impediment standing in the way of international stability and prosperity, but so too was the nation-state more generally. In the past, he had, he confessed, 'too readily' and 'too complacently' accepted the 'existing nation-state, (large and small)' as the basic unit of 'international society'. By the time the war had come to an end he had arrived at a very different answer to his original question about how to organize the new international order. Indeed, if there was to be one at all, it was absolutely vital to think of security less in terms of nations and more in terms of meeting those most basic of human rights such as freedom from fear, from want and from unemployment, rights that transcended boundaries and united peoples

---

[67] See C. A. W. Manning's review of E. H. Carr, *Conditions of Peace* (London: Macmillan, 1942) in *International Affairs*, Vol. XIX, No. 8, June 1942, pp. 443–4.

[68] *Conditions of Peace*, p. xxiii, n. 2.

[69] See Jonathan Haslam, *The Vices of Integrity: E. H. Carr 1892–1982* (London: Verso, 1999), pp. 96–100.

[70] See E. H. Carr, *The Twenty Years' Crisis 1919–1939: an Introduction to the Study of International Relations* (London: Macmillan, 2nd edition, 1946) with a 'Preface To Second Edition' completed by Carr on 15 November 1945. See p. viii.

whatever their nationality. Carr had few illusions about how difficult it would be to arrive at such a 'utopian' solution. Nonetheless, it would in his view be 'an illusion' to think that security could be attained through the vehicle of the nation-state. It would be equally naïve to suppose that the growing and irresistible demand for social justice could be attained by returning to the 'free' international economy of the nineteenth century. If mankind was to learn from the past, it had to face up to the simple truth that nineteenth-century solutions were not enough in the modern age; and if history taught anything, it was that 'old traditions' would 'have to be discarded' and 'new ones created before Europe and the world' could 'recover their balance in the aftermath of the age of nationalism'.[71]

Mr Carr goes to Washington

The publication of *Nationalism and After* marked the beginning of an important transitional period in Carr's already hectic life. In 1945 he signed yet another contract with Macmillan, this time to publish a major study on the History of Soviet Russia, though little anticipating that it would take him over thirty years to complete.[72] A year later he brought out two more books, one a partially (but controversially) revised edition of *The Twenty Years' Crisis* itself, and the other a relatively short and popular study entitled *The Soviet Impact on the Western World*, possibly his most pro-Soviet volume to date which contained what even Carr later agreed were a few 'exaggerations'.[73] In 1947 he then resigned his position at Aberystwyth. Thereafter, he held a few temporary appointments before being elected to a Fellowship at Trinity College Cambridge in 1955. Finally, of course, he was very much present at the creation of a new world order, but unfortunately one which bore little resemblance to that which he had looked forward to for so long. Indeed, as the Cold War intensified, Carr found that his particular style of radical theorizing, combined with what many saw as his pro-Soviet outlook, made him something of a political outsider in the new Britain of the late 1940s and early 1950s. Vilified by his conservative enemies on the right, and attacked quite openly by

---

[71] See E. H. Carr, *Nationalism and After* (London: Macmillan, 1945).

[72] How significant Carr regarded the publication of what he termed – without irony – his 'magnum opus' on *A History of Soviet Russia* can be gleaned from letters sent to his publishers on 12 August 1949 and 21 April 1950 (MAB).

[73] Carr referred to *The Soviet Impact on the Western World* as 'the sort of book which one wants to be got out quickly'. Letter to Macmillan Publishers, 21 May 1946 (CPB).

economic liberals like Hayek (who referred to him as one of the 'totalitarians in our midst')[74] and academic insiders like Isaiah Berlin (who shared his fascination with nineteenth-century Russia but hated his politics),[75] Carr effectively became a prophet outcast in his own country.

However, while all this was happening at home, within the newly established bastion of free world anti-communism, his study on international relations started to exert some considerable influence. *The Twenty Years' Crisis* had already acquired a fairly wide readership in the United States before the end of the Second World War and its reissue in 1946 then made it more widely available. Carr had further contributed to his reputation (and notoriety) through the publication of *Conditions of Peace* and *Nationalism and After*, both of which appeared in American editions. Some insiders were also aware of his many editorials written for the London *Times*, and while a few might have appreciated the writing of this quasi-anonymous dissident at the heart of the English establishment, others took a less charitable view of the steady stream of pieces flowing from Carr's pen attacking the fundamentals of the free enterprise system. A number of Americans were equally sensitive to the fact that Carr's radical views were none too popular amongst certain British politicians either. Indeed, at dinner one evening, the Duke of Devonshire was even reported to have attacked what he and many others viewed as the left-wing drift in *The Times*, and while nobody named Carr, many believed that it was this 'Red Professor from Printing House Square' who was behind the whole thing. This certainly seemed to be the view across the Atlantic where Carr amongst others was seen (by the *New York Times* no less) as having transformed this once solid bastion of the British ruling class into what one critic called the 'the final edition of the *Daily Worker*' – then the daily newspaper of the British Communist Party.[76]

Carr was not an unknown in the United States therefore. Nor was he unfamiliar with the country itself. He had visited America in the first three months of 1938 (he even had tea at the White House with Mrs Roosevelt) and thereafter kept a close watch on developments there as the war unfolded and it became increasingly clear that Britain's leading role in world affairs was gradually giving way to a new *Pax Americana*. Carr

---

[74] F. A. Hayek, *The Road To Serfdom* (London: George Routledge & Sons Ltd. 1944), pp. 138–41.
[75] See Michael Ignatieff, *Isaiah Berlin* (London: Chatto & Windus, 1998), pp. 235–6.
[76] See Charles Jones, '"An Active Danger": E. H. Carr at *The Times*', in Michael Cox (ed.), *E. H. Carr: a Critical Appraisal*, pp. 68–87.

viewed such a trend with a sense of regret and concern, and warned policy-makers of the dangers of drawing too close to the United States – partly because this would reduce Britain's room for manoeuvre in the post-war world, but largely because this would alienate the USSR and lead to a new confrontation which he very much wanted to avoid. Not surprisingly, the almost irresistible train of events that led to the Cold War disturbed him greatly. Carr did not blame the West entirely for this drift. However, there was little question in his mind that the main responsibility for the breakdown in relations was America's refusal after the war – like Britain's before it – to accommodate the rising power of a potential rival.[77]

It was within the context of a conflict that Carr fundamentally opposed, in a country which he largely blamed for having initiated it, that *The Twenty Years' Crisis* had its greatest success in the immediate post-war years. In many ways Carr was a most fortunate author. His volume was one of only a handful on the market, and although written by a non-American had the advantage of being published by someone from a country that still had more than its fair share of admirers in the United States. It also happened along at a particularly important and favourable juncture when international relations as an academic discipline was beginning to take off as a taught subject in American universities. Naturally, there were other books available, and many more would follow as the subject blossomed in the 1950s.[78] However, for the moment, in the land of the rising superpower where numerous courses on 'IR' were beginning to proliferate alongside those of other expanding disciplines such as Sovietology ('know the enemy'), Strategic Studies ('know how to deter the enemy') and Area Studies ('know where the enemy will strike next'), *The Twenty Years' Crisis* proved to be particularly popular. As one American reviewer had predicted, it was always likely to do well and 'would make an admirable introductory text for any college course in diplomatic history or international relations', especially as it provided both an excellent analysis of the causes of war and an even 'more fruitful' discussion of the problems facing the peace-makers.[79]

---

[77] See, for example, E. H. Carr, 'The Russian Revolution and the West', *New Left Review*, Number 111, September–October 1978, pp. 25–36.

[78] For a guide see William C. Olson, 'The Growth of the Discipline', in Brian Porter (ed.), *The Aberystwyth Papers: International Politics 1919–1969* (London: Oxford University Press, 1972), pp. 3–29.

[79] See W. A. Griswold, review of *The Twenty Years' Crisis*, featured as a sales puff in the American edition of the volume which was published in 1964 by Harper & Row Publishers of New York.

The first to discover this, much to his own surprise one suspects, was Carr himself when he returned to America to teach and research in 1948. After having delivered many lectures in 'various universities' across the United States, he found out – much to his chagrin – that while *The Twenty Years' Crisis* was in great demand, his British-based publishers were failing to meet this with a steady supply of books. This not only made little sense from a commercial point of view, it was also very annoying, especially as Carr had discovered at first hand that 'the book was being fairly widely used as a textbook in international relations'. What he had also found out was that 'the subject' was 'increasing' in the US by what he called 'leaps and bounds'. It was a matter of some urgency therefore that Macmillan ship out more copies without any further delay.[80] Two months on and Carr was still urging his publisher to remain ever-vigilant. It was critically important, he felt, to keep the 'American market well fed'. The volume had proven to be a success, 'largely' in the 'universities'. But it was vital to keep stocks up. Sales had already been adversely affected by a failure on the part of Macmillan to 'guarantee copies' in enough numbers in the past, and it was essential not to let this happen again in the future. As he pointed out, American 'professors and tutors had a habit in this country' (he noted with some incredulity) of 'asking publishers in advance whether copies of a book' were available before they would 'put it on "reading lists for students"'. And many, he had heard, had not done so because they could not get hold of it. Carr was most insistent, admitting that while he would not be quite so optimistic about sales of some of his other work in the United States, he was 'sure' that given conditions in America there would be a 'lasting demand' for *The Twenty Years' Crisis*.[81]

Carr's optimism about the potential market for *The Twenty Years' Crisis* in the United States was not misplaced. Its analysis of past failures and rather bleak assessment of what motivated men and nations, seemed to match the American mood as it entered into the dark tunnel that was the Cold War. Naturally enough, not everyone appreciated his materialist epistemology or his debt to Marxism.[82] But what one did not appreciate one could simply ignore. The book, after all, told a riveting story about the illusions of liberalism and the collapse of an international system, and

---

[80] Letter from E. H. Carr to Mr Macmillan, 2 April 1948 (MAB).

[81] Letter from E. H. Carr to Mr Macmillan, 14 June 1948 (MAB).

[82] One of Carr's strongest American defenders (who later helped set up the British Committee on the Theory of International Relations) was nonetheless compelled to criticize his work for its 'Marxist orientation'. See Kenneth Thompson, *Masters of International Thought* (Baton Rouge: Louisiana State University Press, 1980), p. 77.

many read it as a warning to a new generation not to repeat the same mistakes again. This was one of the reasons presumably why Carr's work held a particular fascination for one man in particular: the German émigré, Hans J. Morgenthau. Though he was less radical than Carr as a function of his economics, and more inclined to think of the drive for power in terms of human nature rather than the uneven distribution of resources, there was clearly much that the two men had in common.[83] Both saw the twentieth century as a tragedy.[84] The two together had deep concerns about the relevance of liberalism as a doctrine for the twentieth century. And they had a similar understanding of history and the history of political thought.[85] In their very different ways, they were also highly critical of policy-makers in the West, and while Carr was perhaps the more dissident of the two, Morgenthau always retained a deep suspicion of those who wielded what he wrote about most – namely political power.

It was not insignificant therefore that Morgenthau decided to write an extended review of Carr's work in the first issue of the journal *World Politics* in 1948. In an essay that was as incisive as it was sparkling, he combined both high praise and serious criticism in equal measure. Morgenthau was in little doubt. Carr, he argued, had written four books since 1939 (*The Twenty Years' Crisis, Conditions of Peace, Nationalism and After* and *The Soviet Impact on the Western World*) that taken together represented a contribution of the 'first order' which 'lucidly and brilliantly' exposed the faults of 'contemporary political thought in the Western world'. They were so good in fact that they almost rivalled in depth one of Morgenthau's (and Carr's) intellectual heroes: Reinhold Niebuhr. However, there was a fundamental flaw and there was no point denying it. The basic problem was less historic than 'philosophic'. Carr, he suggested, had done a wonderful job explaining the crisis; where he had

---

[83] See the first edition of Hans J. Morgenthau's classic *Power Among Nations: the Struggle for Power and Peace* (New York: Alfred A. Knopf, 1948). In this Morgenthau recommends readers to look at *The Twenty Years' Crisis* for the light it throws on 'the intellectual problems to which the study of international politics give rise'. At another point he cites the same book as being one of a few that dealt seriously with the role of 'political power' in world affairs. Elsewhere he briefly mentions Carr's views on 'morality', even though he does not happen to agree with them. See pp. 177, 475, 478, 481, 482, 483.

[84] R. B. J. Walker, 'History and Structure in International Relations', in James Der Derian (ed.), *International Theory: Critical Investigations* (Basingstoke: Macmillan, 1995), pp. 332–3.

[85] Compare the similarities between *The Twenty Years' Crisis* with Hans J. Morgenthau, *Scientific Man vs Power Politics* (Chicago: The University of Chicago Press, 1946).

failed was in finding a moral solution to it. Here Morgenthau effectively restated what others had already said before – that Carr's relativism made it impossible for him to develop a 'transcendent point of view from which to survey the political scene and to appraise the phenomenon of power'. The result was to transform Carr the critic and 'political moralist' into Carr the 'utopian of power' whose only means of knowing what was right or wrong was by the criteria of power itself. Hence, whoever held seeming superiority of power became of necessity the repository of a superior morality as well. It was one thing quoting Machiavelli; however for an observer standing outside the world of power, it was still a dangerous thing to be a Machiavelli. It was even worse to be a Machiavelli 'without *virtu*'.[86]

Though critical, Morgenthau's review obviously did Carr's reputation a great deal of good in the United States, where sales of the book continued to be steady. However, there was a limit to how well a complex study originally entitled *Utopia and Reality*, written by a radically-inclined Englishman with a distinct bias against the capitalist system, was ever going to do in the United States. Moreover, as the subject continued to expand so too did demand for more accessible textbooks, and whatever else might be said about *The Twenty Years' Crisis* it was most definitely not that. It may well have been called an 'introduction to the study of international relations', but it could not have read like one to the average American graduate looking for a basic route map through contemporary world problems. This is why Morgenthau's own volume, *Politics Among Nations*, first published in 1948, was to do so well. Brought out by a commercial publisher and written very specifically for the student market, it established itself as *the* book on international relations in ways that *The Twenty Years' Crisis* could never have done. With its simple structure and lists of these principles of realism, those definitions of foreign policy, and its four reasons why disarmament could not work, it was the perfect vehicle for those looking for a quick and easy entry into the subject. By contrast, *The Twenty Years' Crisis* was not only structurally difficult and intellectually more demanding, but did not even begin to address the sorts of issues such as deterrence and American grand strategy, that were fast

---

[86] Hans J. Morgenthau, 'The Political Science of E. H. Carr', *World Politics*, Volume I, No. 1, 1948, pp. 127–34. A shortened version of the same article was later reissued under the more pungent title, 'The Surrender to the Immanence of Power: E. H. Carr' in H. J. Morgenthau (ed.), *Dilemmas of Politics* (Chicago: Chicago University Press, 1962), pp. 350–7.

becoming part of the staple fare in most American courses on international relations.[87]

Carr also suffered as the discipline itself began to evolve in the United States from what it had been in the initial stages – a subject as much influenced by history and political philosophy as by political science – to what it gradually became in the 1960s and 1970s: quantitative, behavioural, and narrowly scientific in scope. In this brave new measured world of game theory, where one's level of analysis appeared to count as much if not more than one's understanding of the past, Carr's volume, with its extraordinary range of references drawn from a host of writers who were either unknown or taboo in the United States, must have made *The Twenty Years' Crisis* look not only outdated and wordy (and possibly subversive too) but beside the point. No doubt historians of the discipline still found him interesting, as did political theorists.[88] But as the focus of the subject altered and the issues it addressed changed, then *The Twenty Years' Crisis* began to look more and more like an historical hangover from a bygone age. Not that Carr disappeared from sight completely. In 1964 his book came out in a popular softback American edition, which did quite well. In 1967 he was then made the subject of a major review, though unfortunately not a particularly good or positive one.[89] And a few years later he resurfaced once again through the work of Robert Gilpin, whose interest in the concept of hegemony and hegemonic stability led him to engage in a fairly serious way with Carr's analysis of the inter-war system.[90] For some at least *The Twenty Years' Crisis* still retained some fascination, even in an age of high positivism when neither his questions nor his answers seemed as pertinent as they once

---

[87] See Chris Brown, *Understanding International Relations* (Basingstoke: Macmillan Press – now Palgrave, 1997; 2nd edn, 2001), pp. 31–2.
[88] See the comment by Daniel H. Deudney, 'Regrounding Realism: Anarchy, Security, and Changing Material Context', *Security Studies*, Vol. 10, No. 1, Autumn 2000, p. 19.
[89] Whittle Johnston, 'E. H. Carr's Theory of International Relations: a Critique', *The Journal of Politics*, Vol. 29, No. 4, November 1967, pp. 861–84.
[90] Robert Gilpin referred to Carr (alongside Thucydides and Machiavelli) as one of the 'three great realist writers'. He later noted that he found in Carr's work 'one of the greatest inspiration for his own scribblings in the field' and 'incorporated Carr's analysis of the relationship of international economic and politics into his own work'. See his 'The Richness of the Tradition of Political Realism' in Robert Keohane (ed.), *Neorealism and its Critics* (New York: Columbia University Press, 1986), pp. 306, 309. On Gilpin's debt to Carr see his *Global Political Economy: Understanding the International Economic Order* (Princeton: Princeton University Press, 2000), pp. 100–1, n.52.

had to an earlier generation growing up in the shadow of the Second World War and the early days of the Cold War.

Carr's growing marginalization from mainstream international relations in the United States in the 1960s and 1970s, should not, though, obscure his very real impact.[91] As yet another American admirer put it, *The Twenty Years' Crisis* was in effect the 'the first "scientific" treatment of modern world politics' and was regarded as such by very many of those who entered the discipline by one route or another in the years immediately following the war.[92] Carr also exerted an influence in another, possibly less positive way: by furnishing students with a particular story about how international relations had evolved and matured into a serious and scientifically respectable academic discipline. The story was simple and optimistic, and like all the best stories it had a happy ending. According to those who told it, this wily Englishman engaged in mortal intellectual combat with a range of woolly-minded thinkers who did not understand the real world, and after several skirmishes met and defeated them in a battle that came to be known as the 'first great debate'. Naturally enough he won the encounter, and consequently helped establish international relations as a proper subject, cleansed of the 'utopian' nonsense that had presumably held it back in the early years. This epic tale of Carr the impaler, confronting and finally defeating the beast of idealism, was repeated so often that few questioned what to later writers looked like a gross simplification. The fact that the beast was never killed, and that the impaler might have set up a series of straw men in the process, was largely ignored. So too was the rather obvious fact that there might have been other writers around who also contributed to the birth of the subject in the modern era.[93] But again, this was politely passed over in silence by those who perhaps needed such simplifying 'foundational myths' to help them on their way.[94] In this

[91] 'Carr' also had 'the distinction', according to a British academic, of 'writing one of the few books on international relations, which despite its subject matter, one might read for pleasure'. See Michael Nicholson, *Formal Theories in International Relations* (Cambridge: Cambridge University Press, 1989), pp. 26–7, n.1.

[92] See Stanley Hoffman, 'An American Social Science: International Relations', *Daedalus*, Vol. 106, No. 3, Summer 1997, pp. 41–60

[93] Daniel Deudney notes that Carr managed to ignore the 'early literature of geopolitics' produced by German writers associated with the Institute of Geopolitics in Munich. See his 'Geopolitics and Change', in Michael Doyle and G. John Ikenberry (eds), *New Thinking in International Relations Theory* (Colorado: Westview Press, 1997), p. 51.

[94] See, for example, Steve Smith, 'International Relations: still an American Social Science?', *British Journal of Politics and International Relations*, Vol. 2, No. 3, October 2000, pp. 376–9.

fashion, though unbeknown to Carr, *The Twenty Years' Crisis* was transformed into an historic morality tale with its wise sages and naïve fools, with the former in the end vanquishing the latter and in the process purging the discipline of its early optimistic excesses.[95]

The other way in which Carr shaped the course of international relations in the United States was more theoretical than mythical, and this was by helping legitimize a particular theoretical approach that went under the generic name of 'realism'. The precise extent to which he did so given the obvious differences between himself and Morgenthau, not to mention later versions of realism which possessed neither his critical understanding of the world nor his sensitivity to historical change, has led several writers to doubt there was ever a relationship at all.[96] However, there was a connection of sorts.[97] Carr may well have been a very different sort of realist, but *The Twenty Years' Crisis* still addressed the fundamental problem of how inequality among states led to conflict between them, which remained of central concern to more mainstream writers.[98] This might not have made Carr 'the realist's realist',[99] any more than his particular brand of realism prevented him from both justifying the appeasement of Hitler while attacking the foundations of bourgeois society at the same time.[100] As has been observed, there are many

---

[95] See Brian C. Schmidt, *The Political Discourse of Anarchy: a Disciplinary History of International Relations.*

[96] On the ways in which Carr was read or misread in the United States see the suggestive comments by James Der Derian, 'A Reinterpretation of Realism' in his *International Theory: Critical Investigations*, pp. 363–96 and Vendulka Kubalkova 'The Twenty Years' Catharsis: E. H. Carr and IR' in Vendulka Kubalkova, Nicholas Onuf and Paul Kowert (eds), *International Relations in a Constructed World* (New York: M. E. Sharpe, 1998), pp. 25–57.

[97] See Joseph Grieco, 'Realist International Theory and the Study of World Politics', in Michael Doyle and G. John Ikenberry (eds), *New Thinking in International Relations Theory* (Colorado: Westview Press, 1997), p. 107. Grieco argues that Carr remains a realist because he argues that 'states coexist in a context of international anarchy' in the 'shadow of war' (p. 107). See also the ways in which Carr's theory of power and peaceful change is employed by Randall L. Schweller and William C. Wohlforth, 'Evaluating Realism in Response to the End of the Cold War', *Security Studies*, Vol. 9, No. 3, Spring 2000, pp. 60–107.

[98] See Graham Evans, 'All States are Equal but...', *Review of International Studies*, Vol. 7, 1981, p. 62.

[99] J. D. B. Miller, 'E. H. Carr: The Realist's Realist', *The National Interest*, Fall 1991, pp. 65–71.

[100] See John Rawls, *The Law of Peoples* (Cambridge: Harvard University Press, 1999), p. 6 n.8, Peter Wilson, 'E. H. Carr: the Revolutionist's Realist', Unpublished Ms, 2000, 16pp, and William C. Olson and A. J. R. Groom, *International Relations: Then and Now: Origins and Trends in Interpretation* (London: Routledge, 1991), p. 93.

different types of realist and various forms of realism.[101] But that has always been the 'trouble' with a theory that in one set of hands can justify the use of power, and in others (like Carr's) can be turned into a critical weapon to attack those who would defend the status quo.[102]

Carr, Merrie England and the 'English School'

One of the many ironies about the history of *The Twenty Years' Crisis* is that while its meaning was being vigorously debated in the United States, in the land of Carr's birth it was revered but rarely discussed at all. Even in his *almer mata* back in Wales it was hardly a cause for fevered discussion. Indeed, according to one who joined the department in 1961 he had 'no recollection of Carr ever being mentioned in lectures or seminars'.[103] Certainly, the more general relationship between the man on the one hand, and what passed for the subject of international relations in Britain in the 1950s and 1960s on the other, could hardly be described as intimate. He reviewed the odd book,[104] and once even read a manuscript by C. A. W. Manning in 1961, though only to warn the publishers that if they were to publish it they had better be prepared for some 'perfunctory and hostile reviews'.[105] He also attended a conference commemorating the fiftieth anniversary of his old department back in Aberstwyth,[106] and on one occasion was persuaded by NATO to give a lecture on the world scene, which he appropriately entitled 'The Twenty Years' Crisis'.[107] But that was about the sum of it. Carr had very little interest in the subject as such, while some of those who purported to represent the subject had, it

[101] See Michael Doyle, 'Peace, Liberty and Democracy: Realists and Liberals Contest a Legacy', in Michael Cox, G. John Ikenberry and Takashi Inoguchi (eds), *American Democracy Promotion: Impulses, Strategies, and Impacts* (Oxford: Oxford University Press, 2000), pp. 23–7, and Michael Joseph Smith, *Realist Thought from Weber to Kissinger* (Baton Rouge: Louisiana State University Press, 1986), p. 2.

[102] Justin Rosenberg, *The Empire of Civil Society: a Critique of the Realist Theory of International Relations* (London: Verso, 1994), pp. 9–37

[103] See Ken Booth, What is His Story? Remarks for the Roundtable, E. H. Carr – The Aberystwyth Years: 1936–1947. Unpublished Manuscript, 13 July 1997. Presented to E. H. Carr: A Critical Reassessment. An International Symposium.

[104] See, for example, his review of Hans J. Morgenthan's book on American foreign policy published in *The Listener*, 15 May 1952.

[105] The manuscript by C. A. W. Manning was *The Nature of International Society* published in 1962 and reissued in 1975. Letter E. H. Carr to Macmillan Publishers, 23 January 1961 (MAB).

[106] See Brian Porter (ed.), *The Aberystwyth Papers: International Politics, 1919–69*.

[107] See Dr E. H. Carr (United Kingdom) 'The Twenty Years' Crisis'. Lecture delivered on 18 July 1957, NATO Unclassified, 18pp (CPB).

seems, little desire to engage in a positive way with him.[108] In fact, when a number of leading lights, including the Cambridge historian, Herbert Butterfield, decided to form what became known as the 'British Committee on the Theory of International Relations' in the late 1950s, they did not even invite him. As one of the initial founders of the group, Martin Wight made clear: 'I hesitate about E. H. Carr' not on personal grounds, but 'because he is so much of a Great Power in this region that I should have misgivings lest he might deflect our discussions into channels opened up by his own work'.[109] There would have been little point inviting him anyway. The central intellectual concept around which the Committee was originally formed – the idea of an 'international society' of states bound by common rules and institutions for the conduct of their relations – was one that was hardly likely to recommend itself to someone as 'Hobbesian' as Carr.[110]

The history of the Committee and its relationship with Carr was by any measure a strange one. Its members obviously knew about his work in general and *The Twenty Years' Crisis* in particular. Martin Wight had even written a penetrating and critical review of *The Twenty Years' Crisis* in 1946 in which he not only chastised Carr's 'provocative' but 'unsatisfying' study, but also claimed that one of his early critics, Leonard Woolf, had penned a 'deadly reply'.[111] The conservative Wight thereafter never appeared to be anything but hostile, and though he felt obliged to come to terms with what even he recognized was a 'brilliant' study, was very much opposed to Carr, and, according to one account at least, 'treated' him 'with disdain'.[112] He was certainly critical of his views and at different times attacked his defence of appeasement, what he saw as his Marxist-

---

[108] The only positive comments that Carr ever uttered about international relations in Britain that I have been able to track down can be found in his 'Academic Questions', *Times Literary Supplement,* 16 April 1954.

[109] The letter was from Martin Wight and is quoted in Tim Dunne, *Inventing International Society: a History of the English School* (Basingstoke: Macmillan Press – now Palgrave, 1998), p. 93.

[110] The term Hobbesian was, of course, used by Martin Wight to describe the 'brilliant' *The Twenty Years' Crisis* in his 'Western Values in International Relations', in Herbert Butterfield and Martin Wight (eds), *Diplomatic Investigations: Essays in the Theory in International Relations* (New York: Columbia University Press, 1966), p. 121.

[111] Martin Wight 'The Realist's Utopia', *Observer,* 21 July 1946, p. 3.

[112] Roger Epp, 'Martin Wight: International Relations as a realm of persuasion', in F. A. Beer and R. Hariman (eds), *Post-realism: the Rhetorical Turn in International Relations* (East Lansing, MI: Michigan State University Press, 1996), p. 122.

influenced theory of ideology,[113] and the fact that he regarded 'power' as being 'anterior to society, law, justice and morality'.[114] Nor was one of the other members of the Committee, Butterfield, especially enamoured of his Cambridge colleague either, and for good reason. Carr had after all attacked him in no uncertain terms in 1951 as one of the many historians who propounded what to his eyes at least seemed to be a reactionary theory that denied the possibility of progress.[115] He had then criticized him again a decade later, and, more robustly still, for having written a 'remarkable book' *(The Whig Interpretation of History)* that was remarkable amongst other things for not having named 'a single Whig except Fox, who was no historian, or a single historian save Acton, who was no Whig'.[116] Butterfield in turn had little time for Carr whose defence of the Russian revolution left the Christian Butterfield very cold indeed.[117] Butterfield was even moved to attack Carr's views expressed in his best-seller *What is History?* published in 1961. While conceding that the book was 'most interesting' he still thought that there was 'a distinct tendency' within Carr's approach which 'operates to undermine the status of history as an autonomous science'.[118] Interestingly, it was through his prompting that Carr in the end was refused an invitation to join the Committee.[119]

The renewed interest in Carr, if it can be called that, did not stem from any work done on him by the Committee or by any attempt by this distinctly exclusive group to come to terms collectively with *The Twenty Years' Crisis*.[120] Rather it manifested itself in one of the more serious individual efforts by one of its members, Hedley Bull, to assess its

---

[113] See Martin Wight, *Power Politics* (Harmondsworth: Penguin Books, 1979), pp. 19, 213.

[114] See Martin Wight, 'An Anatomy of International Thought', *Review of International Studies*, Vol. 13, 1987, p. 222.

[115] See Carr's critique of Herbert Butterfield as an anti-progressive historian in his *The New Society* (London: Macmillan, 1951), pp. 4–9, 16.

[116] For Carr's attacks on Butterfield see E. H. Carr, *What Is History?* (Harmondsworth: Penguin Books, 1961), pp. 19, 41–2, 51, 74, 121.

[117] For a sympathetic assessment of Butterfield see Ian Hall, 'Historical Learning and Historical Thinking: Sir Herbert Butterfield and International Relations', *Review of International Relations*, 2002.

[118] Herbert Butterfield, 'What is History?', *The Cambridge Review*, 2 December 1961.

[119] An important fact recalled by Ian Hall in his 'Still the English patient? Closures and inventions in the English School', *International Affairs*, Vol. 77, No. 3, 2001, p. 504.

[120] Apart from Carr, others not invited to attend the meetings of the British Committee included Harry Hinsley, then lecturer in History at Cambridge, and C. A. W. Manning, Montagu Burton Professor of International Relations at the LSE, 1930–62.

contribution and relevance thirty years after its original publication. An Australian by birth who came to Britain before returning home in the second half of the 1960s, Bull had been taught by both Wight and Manning at the LSE, though his main inspiration initially was Philip Noel-Baker, even though Bull strongly opposed Noel-Baker's support for disarmament. In the 1960s he then turned his attention to the issue of how international order could be maintained under conditions of sovereignty and anarchy, and arrived at the conclusion, via a reading of the seventeenth-century legal theorist Grotius, that states did not inhabit a Hobbesian universe but rather constituted a society of their own bound together by law. At first sight, therefore, there was little to suggest that Bull and Carr had a great deal in common. But Bull did have an interest in *The Twenty Years' Crisis* and felt obliged to deal with it seriously. Originally entitled 'E. H. Carr and the Fifty Years' Crisis', his critique was finally published as '*The Twenty Years' Crisis* Thirty Years On'. In many respects, the paper not only represented 'an important stage in the evolution' of his own thinking,[121] but also constituted something of a manifesto for those who felt that Carr represented more of an obstacle to the development of a mature international relations than an inspiration.

Bull did not reject *The Twenty Years' Crisis* in its entirety. Indeed, there were certain things with which he strongly agreed, including its realist rejection of liberal internationalism and suggestive observations on the problem of peaceful change. Bull even went so far as to suggest that Carr's more general 'analysis of what international politics' was like was 'correct'. However, there were many more things with which he happened to disagree. Most obviously he rejected the book's 'relativism' and by way of example approvingly cited Morgenthau's earlier observation that Carr, lacking any objective perspective of his own from which to make an ethical judgment about the world, was always compelled to 'surrender' to what the American writer had called 'the immanence of power'. It was no accident of course that Carr had been in favour of appeasement, for it flowed logically from what Bull defined as his 'relativist and instrumentalist conception of morals' in which it was impossible to distinguish between good and evil, right and wrong. Bull's other more substantial objection was historical. *The Twenty Years' Crisis*, he believed, was in essence a 'tract for 1939, not for 1969'. It was important therefore to develop a theory of international relations that reflected new realities. According to Bull, when Carr had written his

---

[121] Quote from Kai Alderson and Andrew Hurrell (eds), *Hedley Bull On International Society* (Basingstoke: Macmillan Press – now Palgrave, 2000), p. 125.

volume the world was a very different place. Europe stood at its heart, Wilsonianism remained a genuine force in international politics, realism had yet to assert itself, and the main line of division internationally was that deriving from the legacy of 1919. Thirty years on and the situation had altered dramatically. European domination was a thing of the past, the lessons of realism had been assimilated, the main divisions in the world were no longer between the defenders and the critics of the Versailles system, and the greatest challenge now was not how to deal with 'have-not' powers like Germany but the new revisionist states in the Third World like communist China. In short, new times posed different questions that were bound to lead to different types of answers.

It was at this point that Bull laid his own particular theoretical cards on the table. Carr, he agreed, had done a fine job demolishing old nineteenth-century illusions about the world, and an even more important one showing how foundational notions developed in one era – and unfortunately retained in another – had been one of the main causes of the twenty years' crisis. However, 'having completed this work of demolition' Carr had not then gone on to consider what 'institutions and devices' would be necessary to sustain and maintain 'order' among states under 'present circumstances'. This then allowed Bull to introduce his own master concept that of 'international society', a notion that Carr 'scarcely recognized' in *The Twenty Years' Crisis*. Indeed, according to Bull, in the 'course of demonstrating how appeals to an overriding international society subserve the special interests of the ruling groups of powers' Carr had jettisoned 'the idea of international society itself'. This was his major weakness as a theorist then and why his views had to be repudiated now: not because he was wrong in terms of his understanding of history or even in rejecting the claims of the idealists so-called; but rather because his whole outlook denied the possibility of there ever having been (or there being) an order based upon common interests, common values and by implication common rules and institutions that taken together formed the basis of 'an international society'. This was critical: critical in terms of Carr but critical too in terms of the future of international relations because it was this 'idea' which Carr had rejected so strongly in 1939 that would, he hoped, form the foundation stone for a 'new analysis of the problem of international relations'. What Carr had started but failed to complete we should 'now begin'.

The extent to which Bull presented an entirely accurate picture of Carr's international relations must be open to some doubt.[122] However, the

---

[122] See Peter Wilson, 'The Peculiar Realism of E. H. Carr', *Millennium*, Vol. 30, No. 1, 2001, p. 125.

importance of the piece did not lay in its literal accuracy, but in what it said about Carr's theoretical denial of something which Bull saw as fundamental. On this issue at least Bull could hardly be faulted. *The Twenty Years' Crisis* had grappled with the question of how states related to each other, but it had arrived at rather different conclusions to those which Bull regarded as being central to the renewal of international relations. Carr agreed. As he admitted a few years after *The Twenty Years' Crisis* had been published, he too had once wondered whether there had been (or could ever be) what he called a 'community of nations', but had arrived at the conclusion that under conditions of inequality it was simply out of the question.[123] He was also familiar with the work of Charles Manning whose ideas had had a big impact on people like Bull, and he did not think much of them. In fact at about the same time that the British Committee was setting out on its own particular journey, he was privately deriding Manning's theoretical claims about the 'nature of international society' as being likely to 'encourage illusions'.[124] Nor did he change his mind much thereafter, and several years later confessed to an American academic that he and others had tried many years ago 'to conjure into existence an international society'. However, the whole project had come to nothing for the simple reason that there was no such thing as 'an international society' but instead 'an open club without substantive rules' – a view diametrically opposed to that of Bull's.[125]

Bull's critique of Carr's theory and Carr's rejection of the core notion that formed the basis of what subsequently became known as the English School, appeared to leave little room for doubt. The two had little in common and there was little point in trying to reconcile what in the end could not be reconciled.[126] Of course, those who constituted the school so-called could hardly ignore *The Twenty Years' Crisis*. It was, after all, one

---

[123] See E. H. Carr, *Nationalism and After*, p. 42, n.1.

[124] Manning's book was entitled *The Nature of International Society* and was finally published in 1962 and reissued by Macmillan in 1975. Carr knew and liked Manning and found him and his ideas to be both 'original and stimulating'; but his basic argument about the existence of an 'international society' was quite simply wrong in his view. Carr wrote: 'My fundamental criticism of this thesis is that the international game implied by it is in fact a creation of small and influential groups in the western world during the past 400 years, and there is no reason to suppose that it will survive the decline of these groups.' He also added that Manning wrote in an 'abstract, elusive, hair-splitting style' and that the 'manuscript scarcely' contained 'a proper name from beginning to end'. See letter from E. H. Carr to Macmillan Publishers, 23 January 1961 (CPB).

[125] Cited in letter from E. H. Carr to Stanley Hoffman, 30 September 1977.

[126] See Cecilia Lynch, 'E. H. Carr, International Relations Theory and the Societal Origins of International Legal Norms', *Millennium*, Vol. 23, No. 3, 1994, pp. 589–620.

of the most famous studies on international relations written by a non-American, and anybody entering the discipline in Britain had to come to terms with it. In the end, however, the distance separating Carr the Englishman and this very specific 'English' contribution to the discourse of international relations, was an extraordinarily wide one.[127] On one side stood those who stressed the centrality of shared norms and that which Carr had long ago rejected – the idea that there was an entity known as a society of states; and on the other was Carr himself, who, in spite of a certain tough-minded historical realism that people like Bull and Wight appreciated, conceived of international politics in a very different way. That he stimulated critics like Wight and Bull goes without saying. Indeed, in a crucial way, it is difficult to understand what they had to say without first having read Carr. But is difficult to think of Carr actually anticipating their work.[128] He may have forced others of a very different persuasion to take him seriously. But, as both Bull and Wight seemed to imply, he represented a hurdle that had to be overcome rather than a resource to be used.[129]

Carr goes critical

If Carr's connection to the English School remained decidedly tenuous, one wonders what he would have made of the new interest taken in his work by a group of writers in the 1980s and 1990s. Critical of those who would subsume him under the banner heading of 'realist', but equally chary of being labelled Marxist themselves, this diverse band came to see in Carr a much underutilized resource that was going begging in the last decade of the Cold War. Motivated by either a more general discontent with the conservative character of international relations as a discipline, or a desire to take security studies and political economy in a more radical direction, supporters of what became known as 'critical theory' came to see in Carr someone who could be deployed in a more creative fashion than he had been in the past. Inspired in some cases by the work of the

[127] See Robert Jackson in *American Political Science Review*, Vol. 94, No. 3, September 2000, pp. 763–4.
[128] For an alternative view, see Tim Dunne, *Inventing International Society*, p. 13, and Charles Jones, 'Christianity in the English School: Deleted but not Expunged'. Paper given at the International Studies Association, Annual Conference, Chicago, 21–4 February 2001, p. 3.
[129] The discussion about Carr's relationship to the English School can be followed in the journal *Cooperation and Conflict*, Vol. 35, No. 2, June 2000, pp. 193–237, with contributions by Tonny Brems Knudsen, Samuel M. Makinda, Hidemi Suganami and Tim Dunne.

early German theorist Max Horkheimer who, like Carr, saw a close relationship between the production of ideas and the world of power, and in others by the non-economistic Marxism of the Italian Communist Antonio Gramsci who popularized the key notion of hegemony, the new critical theorists were to go on to make a distinct contribution to the study of international relations, and one of the ways in which they did so was through a re-engagement with Carr's work.

Perhaps the first move in the reclamation of Carr was taken by the Canadian political economist, Robert Cox. An expert on labour economics who had spent many years of his life working in the International Labour Organization in Geneva, Cox was an eclectic but creative thinker who with the less radical (but equally iconoclastic) Susan Strange had helped build a base within international relations for a more critically inclined political economy. Gramscian rather than Marxist in theoretical orientation, Cox had always been a great admirer of Carr. Significantly though it was not just *The Twenty Years' Crisis* that informed his view of Carr (though it played a role) but Carr's work more generally, including his various biographical studies of Bakunin and Herzen, Marx and Dostoevsky, his later work on the Russian revolution, as well as his thoughts on the nature of history. Indeed, what seemed to impress Cox most was Carr's stubborn refusal to be bound by the sort of constrictions that later imposed themselves on academics trained in this or that particular narrow specialization. As Cox observed, Carr not only happened to be well informed but could combine many different levels of analysis, and so paint a picture of a particular period that was not merely more interesting than the average, but historically more sensitive as well. As Cox put it in a short, but deeply revealing autobiographical reflection,

> Carr brought an historical mode of thought to whatever he wrote. He was equally alive to economic, social, cultural, and ideological matters. He studied individuals, especially those whose intellectual influence marked an era; but most of all, he brought all these elements to an understanding of structural change.[130]

Cox's admiration for Carr was thus enormous. He certainly referred to him in his masterwork published in the second half of the 1980s.[131] But it was his hugely influential article of 1981 that probably did more than

---

[130] See Robert W. Cox and Timothy J. Sinclair, *Approaches to World Order* (Cambridge: Cambridge University Press, 1996), p. 27.
[131] Robert W. Cox, *Production Power and World Order: Social Forces in the Making of History* (New York: Columbia University Press, 1987), p. 157.

anything else to raise Carr's profile amongst a new generation of students looking for alternative ways of theorizing international relations. Yet the importance of the piece lay not so much in what Cox said about Carr and history, but in his insistence in drawing a clear line of demarcation between orthodox realism or neo-realism – against which the article was largely directed – and Carr's own very distinctive approach to the study of world politics. Others had hinted at this in the past, and a few of his more sensitive interpreters had already made the point that Carr's form of realism bore little resemblance to that propagated by more mainstream international relations in the United States.[132] However, it was Cox, probably more than any other writer, who dug the deepest channel between the two, and 'rescued' Carr from what he saw as the iron grip that had been imposed upon him by others in the past. Not that Carr had nothing in common with other realists. On the contrary. However, it was his particular brand of realism that inspired Cox. For instead of accepting power as it was, Carr in his view managed to look behind the ideological masks which power wore and reveal its hidden mainsprings. Cox also liked the way in which Carr looked at the world not as a discrete set of realities but as a totality. This was critically important, for it meant that unlike more orthodox realists who made a sharp distinction between the unit of the state and the international system itself, Carr viewed 'reality' wholistically, never thinking for one moment that how states acted or what they did could be discussed separately from the larger international system. Nor did he accept that the relations between states were doomed to repeat themselves. Indeed, as Cox pointed out, the whole thrust of Carr's discussion was to show how changes within states over time had led slowly and irrevocably to the breakdown of the conditions which had made the nineteenth century possible and the twentieth century so unstable. This again was crucial. In fact, it was only by employing a similar methodology (one which Carr, by the way, had developed quite independently of anything he may have said directly about either international relations or the state) that one would be able to illuminate the evolution of the modern state and global relationships in a very rapidly transforming world where production within nations was giving way to the internationalization of production, where the state as constituted since the Treaty of Westphalia was fast being internationalized itself, and where the forms of hegemony that had formed the basis of world order in the post-war period – a question that Carr had addressed

---

[132] See, in particular, Graham Evans, 'E. H. Carr and International Relations', *British Journal of International Studies*, Vol. 1, No. 2, 1975, pp. 77–97.

briefly in his own study on the inter-war system – were fast dissolving and opening up new historical possibilities.[133]

In his own work Cox himself distinguished between two ways of thinking – one he termed 'problem solving' and the other 'critical'. The former, which was distinctly not Carr, effectively accepted reality as it was and then attempted to solve any problems facing the world within a specific and set framework defined by those who controlled the reins of power. The latter stood apart from the prevailing order and asked how that order had come about, how it had changed over time, and ultimately how it might be transcended. Cox also insisted (and here he almost paraphrased Carr exactly) that all forms of theory must always be for someone or some purpose, and that unless it proposed a way forward it was little more than useless. Here again Carr served as a model thinker who not only discussed the causes of things – in this case the collapse of the inter-war system – but also tried to propose solutions. He did this initially and tentatively in the last few pages of *The Twenty Years' Crisis*, and then in more detail in *Conditions of the Peace*, *Nationalism and After*, *The Soviet Impact on the Western World* and his important study, aptly and significantly called *The New Society*. This forward-looking or even utopian aspect of Carr's work (largely ignored by international relations scholars in the United States and Britain during the height of the Cold War) was one that interested a number of writers other than Cox. One was the political theorist, Andrew Linklater. However, whereas Cox used Carr to inform his work on international political economy and large-scale historical change, Linklater deployed him to think creatively about the possibility of constructing a new theory of political community based upon different notions of citizenship.[134]

Like many in the international relations community, Linklater had for many years regarded Carr's work as being that of an 'unadulterated realist'. But in the process of rethinking his own ideas that would in the end take him 'beyond realism and Marxism' towards a more serious engagement with the idea of 'the transformation of political community', he was forced to come to terms once again with Carr's contribution.[135] He

---

[133]Robert W. Cox, *Production, Power and World Order*, pp. 49–50, 86, 91, 100, 117, 169, 400, 502.

[134] Andrew Linklater, 'The Transformation of Political Community: E. H. Carr and Critical Theory and International Relations', *Review of International Studies*, Vol. 23, 1997, pp. 321–8.

[135] Andrew Linklater, *Beyond Realism and Marxism: Critical Theory and International Relations* (London: Macmillan, 1990). See also his later *The Transformation of Political Community: Ethical Foundations of the post-Westphalian Era* (Cambridge: Polity Press, 1998).

did so first in 1990 in a study that rather tentatively suggested that in *The Twenty Years' Crisis* Carr had tried to resolve the basic 'antinomy between realism and idealism' by setting forth a 'defence of national policy which aimed at the extension of moral obligation and the enlargement of political community'. But the idea was never fully developed and it was to take some time before Linklater would exploit Carr to the full. Interestingly this occurred not through any new reading of *The Twenty Years' Crisis*, but through a discovery of Carr's later work on nationalism and the nation-state. Indeed, it was to be Linklater's meeting with this 'other' Carr, and in particular what he later referred to as his 'magisterial essay' *Nationalism and After* published in 1945, that seemed to inspire him far more than Carr's earlier classic. Linklater's praise for Carr's 1945 study seemed to know no bounds. It was in his view 'an exemplary attempt, unequalled in the field, to show how the achievements of modern political life can be secured while the propensity for violence and exclusion is overcome'; and Carr did so, basically, by both reconfiguring the relationship between the nation-state and the international system on the one hand, and the concept of citizenship and what citizenship meant in a new Europe on the other. The result was one of the more exciting and radical attempts in modern political theory to think through the lessons of the past in order to be able to build a more enlightened future that 'would be more internationalist' than its 'predecessors, more sensitive to cultural differences, and more passionately committed to ending social and economic inequalities'.

What Linklater referred to as his 'morally charged' reading of Carr was not unconnected, of course, to a larger political and intellectual agenda in which he had by now become a key figure. Radical in outlook but open to many influences, the central concern of the new wave was to develop new perspectives in international relations that would challenge inequality and exclusion and in this way advance a form of emancipatory politics that was not just 'problem solving' (to use Cox's terminology) but critical all the way down. Carr, he believed, contributed to this project in at least two important ways: through an embryonic but increasingly sophisticated normative analysis of how the nation-state ought to evolve; and by engaging in a serious (though largely unfinished) discussion of how the study of international relations might be reformed in order to tackle the dominant moral and political questions of the epoch. According to Linklater, this did not make Carr a utopian. As he noted, his proposals for a new politics was not spun out of thin air but was rather immanent in his analysis. On the other hand, it did not suffer from the 'sterility' normally associated with realism. And it was vitally important to reclaim this aspect of his thought, which had hitherto been obscured by his reputation as a realist. Only when this had been done would it be possible to develop a mature international relations which transcended the immediate without descending into fantasy.

Linklater's long march back to Carr was paralleled, at least in part, by the journey travelled by an equally influential writer, Ken Booth. A strategist who had made his reputation in the 1970s, Booth gradually came to question most if not all of the central claims made by realism in the 1970s and 1980s. There was no 'Road to Damascus' conversion, but instead a series of engagements which over time led him to question that which he had once assumed to be true. The first realist truth to go was probably the notion that all states behaved similarly within a given set of determined structures; the next was the idea of the 'Soviet threat' and the logic or illogic of nuclear deterrence; and the last is what he later called the 'Cold Wars of the mind' which supported such policies. In the process, Booth carved out a quite unique position within and outside the international relations community, one that preceded the Second Cold War with the publication of his *Strategy and Ethnocentrism* in 1979, but was then consolidated in the turbulent days which followed in the ongoing debates about the role of nuclear weapons in a divided Europe. An early advocate of the notion of non-offensive defence, and a major enthusiast of Gorbachev's new thinking about foreign policy and European security, Booth not only challenged conventional thinking about security policy but in the process effectively helped to popularize a set of fairly complex ideas that went under the general heading of 'critical security studies'. Premised on the assumption that realism in its more developed form was inadequate, but that any prospect for change had to begin from where we were (rather than where we were not), 'CSS' as it became known ploughed a path somewhere between more radical critics on the far left who dreamed dreams that could never be realized, and the tired complacency that had for so long characterized thinking about security in the West.[136]

It was thus of no little significance that Booth decided to make Carr the subject of his inaugural lecture in 1991. Nor was it accidental. Booth, after all, had come to reject the realist norms in which he had originally been trained, and in Carr he saw a writer who had also grappled in the past with the questions posed by realism and found their answers to be wanting too. He was equally interested, however, in Carr the 'utopian' who though labelled 'realist' had (like himself) pursued what he called the same goal of

[136] See Ken Booth, 'Security and Self: Reflections of Falllen Realist', in Keith Krause and Michael C. Williams (eds), *Critical Security Studies* (Minneapolis: University of Minnesota Press, 1997) pp. 83–120.

trying to reconcile the pursuit of a different world – termed 'utopia' – but had sought to embed this within a given set of facts called 'reality'. Booth called this approach 'utopian realism', which, though not without its problems, appealed precisely because it tried to bring together the two 'planes' of utopia and reality which Carr had said could never meet.[137] Others had pursued the same goals of trying to reconcile the two, but it was possibly Carr more than anybody else who had grappled most consistently with the ever-present issue of seeking to develop a theory of the world based upon the world as an empirical reality but never accepting the world as it was. Not that Carr was without his problems. According to Booth, Carr was rather confused as to where he stood in relation to utopianism and realism. His attacks on the utopians so-called had also done a great deal of damage to the reputation of Kant. However, the important task was not to dwell on the limits of *The Twenty Years' Crisis*, but instead to rescue its many insights from the realists pure and simple. His more orthodox readers in the past had only seemed to notice where Carr the realist was attacking the utopians. It was equally vital, if not more so, to note his uncertainty towards realism as well as where he made the most positive comments about utopianism itself.[138]

The attempt by a number of leading writers to retrieve Carr from realism and exploit his critical potential represented a significant move in international relations. Nonetheless, it was not without its problems. First, there was the not insignificant issue of Carr's economics. Carr lived in an era during which the most readily available answer to the failure of the old economic system was some form of planning that would transcend the limits of the market and make a more effective use of resources in a more equitable way. The events of 1989, followed two years later by the disintegration of the Soviet Union, seemed to put pay to such ideas, and in the process removed at least one of the props supporting Carr's larger vision of a new international order. His conception of that order moreover left a few things to be desired. A modernist in every respect who was ever keen to be on the side of what he saw as 'progress', he was perhaps a little less sensitive than he might have been about the costs which progress tended to exact from those who stood in the way. Indeed, those who did stand in the path of history got short shrift from

[137] For an earlier discussion by Booth of Carr who recognized that all 'sound political thought must be based on elements of reality and utopia' see his '1931 + 50: It's Later than we Think', in David R. Jones (ed.) *Soviet Armed Forces Review*, Vol. 5 (Gulf Breeze FL: Academic International Press, 1981), pp. 31–49.

[138] Ken Booth, 'Security in Anarchy: Utopian Realism in Theory and Practice', *International Affairs*, Vol. 67, No. 3, 1991, pp. 527–45.

Carr (small nations and peasants in particular) while those who appeared to be the bearers of a new age, the big powers, Lenin and even Stalin – that great 'westernizer', as he once referred to him – were more often than not treated with kid gloves. Carr, as Isaiah Berlin once observed, always seemed to be with the big battalions, the winners of history, and not its losers.[139] Finally, if we are to judge Carr as a critical theorist, we can hardly ignore the fact that he effectively provided a rationalization for the foreign policies of Germany before the war, and Soviet Russia after it. No doubt both these 'have-not' states had suffered much at the hands of the more powerful democracies, and Carr was honest enough to say so. Unfortunately, in the process, he ended up by providing a theoretical gloss for what the two powers then did in Eastern Europe. This not only denied certain rights to countries like Czechoslovakia and Poland, but also laid Carr open to the quite legitimate charge of being consistently opposed to the West but not to the West's enemies. Carr might have been critical, but evidently he was not consistently critical enough. To paraphrase a term later made popular in international relations, he was never quite critical 'all the way down'.

Conclusion

As this extended essay has tried to show, *The Twenty Years' Crisis* was not the only book that Carr wrote on international relations; however, it was the one volume that secured his reputation in the field. In a way this was most fortunate for Carr, for whereas what he regarded as his most important contribution to human knowledge – his history of early Soviet Russia – is now hardly read at all, *The Twenty Years' Crisis*, in which he displayed only a limited degree of interest, continues to be read and discussed. Indeed, whereas his reputation as an historian of the USSR waned considerably after the system's collapse in 1991, his standing in international relations appears to have gone from strength to strength. Moreover, this happened during a decade which was not especially kind to more conventional realists. Carr, on the other hand, was never more popular. In 1998 he was the subject of two major reassessments.[140] The following year he got the biography he deserved.[141] And in 2000 there was

---

[139] Isaiah Berlin, 'Mr Carr's Big Batallions', *New Statesman*, 5 January 1962.
[140] Charles Jones, *E. H. Carr and International Relations: a Duty to Lie* and Tim Dunne, Michael Cox and Ken Booth (eds), The *Eighty Years' Crisis: International Relations 1919–1999*.
[141] Jonathan Haslam, *The Vices of Integrity: E. H. Carr, 1892–1982*.

an attempt by scholars from a variety of disciplines to deal in a fair but not uncritical way with his contribution to intellectual life over a half-century period.[142] So much got to be written about Carr in fact that one commentator was even moved to talk of a new 'Carr' industry in the making.[143]

This renewed interest in Carr raises a fairly obvious question about the contemporary importance and relevance of *The Twenty Years' Crisis*. Carr himself always appeared to be in two minds about the book's long-term prospects. He accepted that it might be of some 'interest and significance' to those who read it later. On the other hand, he did not have the confidence of someone like Thucydides the Greek historian, who predicted (correctly as it turned out) that his volume on the Peloponnesian Wars would last 'for ever'.[144] Conceived in a pre-superpower age of well-defined nation-states and national economies, *The Twenty Years' Crisis*, according to Carr just a few years after it had been published, had 'already' become something of 'a period piece'. Several decades on, and that harsh judgement passed in 1945, when it was perfectly normal to assume (as did Carr) that western capitalism had a past but no future, and that planning and progress were synonymous terms, would appear to be more true than ever.

Yet serious claims can still be made on behalf of the book – the most obvious perhaps being that it does provide us with an extraordinarily exciting analysis of the inter-war period; and, as Carr no doubt anticipated, it was a period that was likely to cast a very long shadow. Certainly, the lessons later drawn from these turbulent years, especially the so-called lessons of 'Munich', played a critical role in shaping the way in which subsequent policy makers in the West came to look at the world around them, particularly during the chill days of the Cold War, when many assumed that the only way of preventing another war now was by hanging tough with the Russians in ways that the democracies had not hung tough with the Germans before 1939. From this perspective, Carr served two entirely different masters in the post-war period: one that pointed to his advocacy of appeasement and cried 'never again', and another that relished his attacks on the idealists and applauded him loud and long for reminding the naïve that in an age of aggressive totalitarian

---

[142] Michael Cox (ed.), *E. H. Carr: a Critical Appraisal* (Basingstoke: Macmilan Press – now Palgrave, 2000).
[143] Duncan S. A. Bell, 'International Relations: the Dawn of a Historiographic Turn?', *British Journal of Politics and International Relations*, Vol. 3, No. 1, April 2000, p. 122.
[144] Thucydides, *History of the Peloponnesian Wars* (New York: Penguin, 1972), I. 22.

powers, it was always wise to carry a big stick – even if one did not always have to use it.

Another important reason for revisiting *The Twenty Years' Crisis* is that it is one of those books that can be read on several different levels. Written at a time when one era was about to break down, it can be studied, and more often than not has been, as a particular type of analysis of a very peculiar moment in time. But locked away within its more specific observations about the inter-war system, there was something of a more general character struggling to be heard: namely a theory of international crisis. Eschewing any single-factor structural explanation as to why wars occur or systems break down, Carr in effect identifies three critical factors that have to be present to make any potential crisis dangerous and threatening: the existence of powerful and resentful states situated outside of the international order; a profound and sustained disruption to the operation of the global economy; and finally the unwillingness or inability by any single power or 'hegemon' to underwrite international order. Carr may not have intended it, but situated within his explanation of one 'twenty years' crisis' was a larger thesis about global stability and instability, and how and why international systems in general can and do break down, with the most devastating consequences.

The Twenty Years' Crisis also addresses another major issue of more general concern – that is, how should victorious states set about building and maintaining peace after extended periods of war? The question has been and remains an important one, more so than ever perhaps since the end of the Cold War, when the victors once again were compelled to grapple with what one analyst has called 'the fundamental problem of international relations'.[145] It was certainly a problem with which Carr was familiar. In fact, one of the reasons for originally writing *The Twenty Years' Crisis* was not to reflect endlessly on what had gone wrong in the past, but to ensure that the same mistakes were not made again in the future by those to whom Carr dedicated the book: the 'makers of the coming peace'. And what the past taught was the following. That a successful peace required not only the formation of a set of international institutions but also the creation of a set of social and economic conditions as well; that there could be no peace in Europe unless Germany could find a secure place within it without having to resort to expansion; and that the nation-state had to be regarded as an obstacle to an orderly peace, rather than the vehicle through which order would be achieved. On all counts Carr

---

[145] G. John Ikenberry, *After Victory: Institutions, Strategic Restraint, and the Rebuilding of Order After Major Wars* (Princeton: Princeton University Press, 2000), p. 3.

turned out to be extraordinarily prescient. Indeed, as any study of the way in which those made the peace in 1945 and then again in 1989 would show, indirectly at least they owed a great deal to Carr, and people like Carr, who, having witnessed the failure of one peace settlement, were determined that the same would never happen again.[146]

The Twenty Years' Crisis also has a great deal to say about the power of ideas in general, and the meaning of liberalism in particular.[147] For Carr, of course, liberalism constituted an ideology of the nineteenth century that was bound to be superseded in the twentieth. In this he turned out to be the prophet of a false collectivist dawn. Liberalism not only survived under the umbrella of American power in the post-war period, but its central tenets proved to be far more resilient than he could ever have anticipated. In fact, with the unexpected victory of the West over communism in 1989, it actually experienced a surge of self-confidence that led at least one well-known writer to proclaim that there was now no serious alternative.[148] Yet, as Carr would have observed, what constitutes the common sense of one age can very easily become the dangerous illusions of another, and a decade or so after the victory of liberalism its limits have become only too apparent, even to those who once championed its values of unfettered markets and open economies. And as we move from one era of romantic market triumphalism to another where an increasing number of actors are questioning the limits of the free enterprise system, we could do a lot worse than revisit what Carr had to say on the subject in *The Twenty Years' Crisis*.

Finally, Carr's whole approach to international relations, which understands the present as history and power as a problem that has to be taken seriously, is one that many modern students of the subject could do well to ponder. More recent trends in international relations have led some of its leading practitioners into some fairly arcane corners where discussions about what we know has been replaced by endless debates about whether we can know anything, and where the study of history seems to have been replaced by a series of constructed narratives that take as their point of departure the idea that history as such does not exist. Innovation is no doubt the spice of intellectual life, and all advances in knowledge are likely to be built on the bones of those who have gone

---

[146] See Ian Clark, *The Post-Cold War Order: the Spoils of Peace* (Oxford: Oxford University Press, 2000).

[147] See Alexander Wendt, *The Social Theory of International Politics* (Cambridge: Cambridge University Press, 1999), pp. 195–6.

[148] Francis Fukuyama, 'The End of History', *The National Interest*, Vol. 16, Summer 1989, pp. 3–18.

before. But an international relations that denies the notion of a past, or takes no interest in power, is hardly likely to come up with any major discoveries. This is why *The Twenty Years' Crisis* retains its vitality, as both a reminder of what we need to be studying and why, and why an engagement with history and power does not have to lead to a worship of either. If we take nothing more than that from the study of his classic work, then we may have taken a very great deal indeed.

A Brief Guide to the Writings of E. H. Carr

E. H. Carr wrote so much on so many diverse subjects, over such a long period of time, and for so many different audiences, that it is not at all easy to situate him within any particular academic discipline. Indeed, in some ways it is not even accurate to call him an academic, given his many careers and interests; nor to be true to Carr's own wishes should we even refer to him as a member of the 'IR' community. In fact, he did not much like the idea of international relations as a subject and said so on numerous occasions. Yet it is as the author of one of its classical foundational texts that he is perhaps best known today.

So how should we read Carr and make sense of *The Twenty Years' Crisis*? One very obvious way of doing so is to get some idea of what he wrote and what he did over his whole life, spanning the period from 1916 – when he took up his position in the Foreign Office – to 1982 when he died at the age of 90. Indeed, it is crucial to do so, for only by setting his famous study within his larger *oeuvre* can we fully appreciate the true significance of *The Twenty Years' Crisis* as one of several attempts made by Carr to make sense of a world turned upside down by one war and the threat and reality of another, the collapse of old economic certainties and the challenge posed by the Soviet Union – not in any military sense but as an ideological alternative. Martin Wight once observed that Carr's *The Twenty Years' Crisis* was the product of diseased times. It would be more accurate to say that all of his work is a response to truly revolutionary times.

The first 20 years of his active professional life, however, were spent within the cloistered confines of an elite British Foreign Office. From this privileged, and at times influential vantage point he studied and advised successive governments about the Russian revolution, the Versailles Peace Treaty, the formation and evolution of the League of Nations, the world depression, the rise of Hitler, and how Britain in the early stages of decline – as he perceived it – ought to respond to the threat posed by Germany in the 1930s. The last issue in particular was of special concern to Carr. A close confidante to those within the British foreign policy establishment who had been keen to pursue a policy of détente with Hitler, he saw *The Twenty Years' Crisis* not as an attempt to provide further justification for a strategy that had manifestly failed by the time the book was first published in November 1939 – the standard view – but as a retrospective critique of those who had palpably failed to do so before 1938. Certainly, without recognizing the very special relationship Carr enjoyed with at least some of those who wielded power and made policy, it is virtually impossible to understand how and why he wrote *The Twenty Years' Crisis*, a work which was at once a fairly abstract discourse on the nature of international relations – his preferred title for the book was *Utopia and Reality* – and also a very direct intervention (admittedly after the policy of appeasement had been abandoned) into the key international debates of the day.

Carr, though, was no ordinary diplomat, and while still serving in the Foreign Office he became increasingly interested in Russian literary and revolutionary politics in general, and Karl Marx in particular – not exactly the normal topics of conversation within the British diplomatic corps. These formed the basis of four biographical studies written in the space of just six years, between 1931 and 1937, a remarkable output by any measure. These included *Dostoevsky 1821–1881* (1931; 1949; London: Unwin, 1962), a political and psychological portrait of the great Russian novelist; *The Romantic Exiles* (1933; London: Serif, 1998), which examined the wonderful and bizarre world of Russian revolutionary exiles in Europe in the mid-nineteenth century; *Karl Marx: a Story in Fanaticism* (London: J. M. Dent, 1934), in which Carr looked critically at the life and works of what he called 'this orderly German'; and *Michael Bakunin* (London: Macmillan Press – now Palgrave, 1937), a fascinating study of the Russian anarchist. It was not insignificant that Carr chose not to write about conventional figures, but rather those whom he later described as 'living outside the charmed circle'. But it was not the scientific Marx who influenced him most (though he did influence Carr a good deal more than later commentators have recognized), but his immersion in the world of Russian nineteenth-century intellectuals, even those 'who were not in any strict sense revolutionaries at all'. As he later confessed, it was when reading these writers and thinking about their world – one that was so different to his own – that he began to reflect critically about the 'liberal moralistic ideology' that had so far formed the ideological bedrock of his own outlook.

Carr thus left the Foreign Office in 1936 a very different person than when he had joined it 20 years earlier. But in the end, however, it was not so much his sympathy for Russian intellectuals but his desire to speak more freely about what he saw as the drift in British foreign policy that led him to resign from the Foreign Office in early 1936 and take up the Woodrow Wilson Chair in International Politics in Aberystwyth. Here he remained on and off until 1947, in the process writing several books, six of which were published by Macmillan: the biography of *Michael Bakunin* (1937), which was not a good seller; his textbook on the inter-war period, *International Relations Since the Peace Treaties* (1937; after 1947 *International Relations Between The Wars, 1919–1939*), which sold well; *The Twenty Years' Crisis* (1939); *Conditions of Peace* (1942), perhaps his most trenchant statement about the world crisis, which also contains a powerful critique of liberal political economy; *Nationalism and After* (1945), in which he put forward his most radical thoughts to date on international security without the nation-state; and *The Soviet Impact on the Western World* (1946), which despite its pro-Soviet leanings (Stalin, he claimed, had placed 'democracy at the forefront of allied war aims') did contain some important insights about the ways in which the West had evolved since 1917 in direct response to the challenge posed by the USSR. His other works included his largely ignored but significant 1939 study entitled *Britain: a Study of Foreign Policy from the Versailles Peace Treaty to the Outbreak of War* (London: Longmans, Green & Co, 1939), and his early statement on the failure of the nation-state, *The Future of Nations: Independence or Interdependence?* (London: Kegan Paul, 1941). Three other studies also deserve brief mention here, all of which were composed after he had resigned from Aberystwyth: *Studies in Revolution* (1950), a series of 14 essays, most of which had first appeared in the *Times Literary Supplement; The New Society* (1951), in which he declared his faith in progress and defined freedom as 'the opportunity for creative activity'; and *German–Soviet Relations Between the Wars, 1919–1939* (1952), a standard diplomatic account of a crucial relationship.

As this prodigious output might suggest, Carr was not an average academic. The Wilson Chair moreover was not the standard academic posting and it left him free most of the time to pursue his other interests as a journalist, reviewer and highly active member of various discussion groups at Chatham House in London. Through this latter vehicle he also helped clarify his own ideas on two especially important problems – peaceful change and nationalism – both of which figure prominently in the final draft of *The Twenty Years' Crisis*. He also managed to maintain very close contacts with those who remained at the heart of the British political class. It came as no great surprise, therefore, when he was asked to join the Ministry of Information in 1940 in order to be able to play his part in the war he had hoped Britain might have avoided (at least until 1938) by appeasing Germany. However, realizing that the post did not give him a sufficiently important platform from which to express his views, he decided to take up a position as Assistant Editor and leader writer with *The*

Times. This was a critical move, and from this largely anonymous position he set out to shape the debates about the future of Britain and its position within the larger post-war world. Broadly progressive in politics with little faith left in the private enterprise system, Carr inevitably upset Conservatives on the Right. What upset them more, perhaps, was his continued advocacy of good relations with the Soviet Union, even after the war had come to an end and given way to a Cold War for which he largely (though not completely) blamed the West.

In 1945 Carr took a decision that was to determine the course of the rest of his life. Convinced now that the three most important events of the twentieth century were the Russian revolution of 1917, its entry into the war in 1941, and its emergence as some sort of 'superpower' by 1945, Carr turned his attention away from the world of international politics to write what he hoped would become the definitive history of the early Soviet Union. The final result was to be ten volumes spanning the period 1917–1929. Three dealt with the early years – *The Bolshevik Revolution 1917–1923* (1950, 1952, 1953), one with the transition from Lenin to Stalin – *The Interregnum 1923–1924* (1954), three with the middle period of the 1920s – *Socialism In One Country, 1924–1926* (1958, 1959, 1964), and another three with the new system in formation – *The Foundations of a Planned Economy, 1926–1929* (1969, 1971 and 1978), two of which he co-authored with R. W. Davies (all London: Macmillan Press – now Palgrave). He also published three related volumes: *The Russian Revolution from Lenin to Stalin, 1917–1929* (1979), *The Twilight of the Comintern, 1930–1935* (1982), and *The Comintern and the Spanish Civil War* (1983) (all London: Macmillan Press – now Palgrave). Olympian was the word most frequently used to describe the achievement, and Olympian it most certainly was, especially when we consider that the first of these many volumes was not published until 1950 when Carr was already 58. Little wonder that even he at times felt that the whole project had got completely out of hand.

If Carr had ever been asked what his most important contribution to human knowledge had been, he certainly would not have replied *The Twenty Years' Crisis*. Nor would he even have included his hugely influential but relatively short book on the study of history, *What Is History?* (London: Macmillan Press – now Palgrave, 1961), which went on to sell over a quarter of a million copies in the English language, as well as being translated into 25 others, including Russian and Georgian in 1998, Lithuanian and Korean in 2000, and Macedonian in 2001. First, he would have mentioned his magnum opus on Soviet Russia, a study that dealt in minute detail with the ways in which economic and international realities in the end modified and altered the high ideals of its revolutionary founders. Second, he would have cited (and did in 1980) his now out-of-print and almost impossible to obtain biography of the anarchist Michael Bakunin – a utopian dreamer if ever there was one, whose love of liberty if not understanding of the 'real' world inspired Carr to write about him with such great sympathy.

About *The Twenty Years' Crisis* he seemed to have no strong opinions at all, even though the complex relationship between the world as it is and the world as it might be – or, as Carr might have put it, between reality and utopia – forms the philosophical foundations upon which it is constructed. But Carr never appeared to see it as anything but a particular type of book written under very peculiar and extreme circumstances. As he noted in the preface to the 1981 printing of the volume, by the end of the war *The Twenty Years' Crisis* 'was already a period piece and such it remains'. It might 'perhaps retain a wider significance', he continued, but one should never forget that it was first and foremost 'an attempt to navigate then relatively unchartered waters' and should be judged in those terms. Whether or not he succeeded in his navigational efforts remains an open question. *The Twenty Years' Crisis*, after all, was never an easy book, nor did it ever lack for critics. Its influence, however, can hardly be denied. Reprinted four times within six months of its publication in late 1939, it went through no less than 17 reprintings between 1946 and 1984. By 2000 it had also been translated into seven languages, including Japanese, German, Portuguese, Greek, Swedish, Spanish and Korean, thus making it one of the best known books on international relations in the world.

A Guide to the Secondary Literature on E. H. Carr

The appearance of *The Twenty Years' Crisis* in late 1939 witnessed the start of a long and often acrimonious debate about the book – one that continues today and seems to show no sign of abating. The starting point for an understanding of how Carr was initially read and received has to be the two important essays by Peter Wilson, 'The Myth of the First Great Debate', in Tim Dunne, Michael Cox and Ken Booth (eds), *The Eighty Years' Crisis: International Relations 1919–1939* (Cambridge: Cambridge University Press, 1998), and 'Carr and his Early Critics: Responses to *The Twenty Years' Crisis*, 1939–1946', in Michael Cox (ed.), *E. H. Carr: a Critical Appraisal* (Basingstoke: Palgrave, 2000). Both books also contain much material on Carr himself. The various contributors in the former publication attempt to deploy some of Carr's main categories to see whether or not they have much value for the study of the modern world system – and generally conclude that they do. The second study includes 15 essays that discuss the various facets of Carr's contribution to intellectual life, with a number – those by Brian Porter, Tim Dunne, Andrew Linklater, Paul Rich, Randall Germain and Fred Halliday – dealing with Carr's more specific contribution to international relations. The latter also contains the indispensable autobiographical note by Carr himself. Written only a couple of years before his own death in 1982, Carr explains 40 years after the first appearance of *The Twenty Years' Crisis* that it was 'not exactly a Marxist work' but was nonetheless 'strongly impregnated with Marxist ways of thinking, applied to international affairs'. He also confessed that he was a 'bit ashamed' of its 'harsh realism', and by way of compensation went on in 1940 and 1941 to write what he called 'the highly utopian *Conditions of Peace* – a sort of liberal Utopia mixed with a little socialism but very little Marxism'. For essential background on how and why Carr came to write *The Twenty Years' Crisis* (and a lot more besides), see Charles Jones' somewhat oddly titled, but very useful study *E. H. Carr and International Relations: a Duty to Lie* (Cambridge: Cambridge University Press, 1998), which examines Carr's ideas on world affairs between 1936 and 1946. Rather strangely titled too is the equally impressive biography of Carr written by Jonathan Haslam (a former student but not an acolyte of Carr), *The Vices of Integrity: E. H. Carr,*

1892–1982 (London: Verso, 1999). Two other works should also be consulted, more for what they have to say about the various 'idealists' whom Carr attacked than for Carr himself. They are David Long and Peter Wilson (eds), *Thinkers of The Twenty Years' Crisis: Inter-War Idealism Reassessed* (Oxford: Clarendon Press, 1995) and Lucian Ashworth, *Creating International Studies* (London: Ashgate, 1998).

As Wilson shows, Carr's study provoked something close to outrage among those whom he had either attacked directly or whose ideas he had disparaged. A number felt compelled to reply in kind to what they saw as a mischievous but potentially influential volume with little moral compass, and which came all too close, in their view, of suggesting that high ideas about peace, order and international law were little more than the ideologies of the powerful states seeking to defend the status quo against the 'have-not states' such as Nazi Germany. Significantly, one of the more significant contributions came from the noted liberal Norman Angell in his *Why Freedom Matters* (Harmondsworth: Penguin Books, 1940). Leonard Woolf also replied in his *The War for Peace* (London: Routledge, 1940) and 'Utopia and Reality', *Political Quarterly*, Vol. 11, No. 2, April–June 1940, as did Susan Stebbing in *Ideals and Illusions* (London: Watts and Co., 1941). His predecessor at Aberystwyth and first holder of the Woodrow Wilson Chair, Alfred Zimmern, also criticized Carr in 'A Realist in Search of Utopia', *Spectator*, 24 November 1939; so too did R. W. Seton-Watson in 'Politics and Power', *Listener*, 7 December 1939. See though the more positive reviews by two American writers, W. P. Maddox in *American Political Science Review*, Vol. 34, No. 3, 1940, pp. 587–8, and Crane Brinton, 'Power and Morality in Foreign Policies', *Saturday Review*, 17 February 1940. However, it was left to the British socialist Richard Crossman to make what many thought was the most damaging political point of all: that in his rush to demystify power in international relations, Carr ended up writing an apologia for the foreign policy of Germany in the late 1930s; see his 'The Illusions of Power', *New Statesman*, Vol. 28, No. 475, 25 November 1940, pp. 761–2. An anonymous reviewer of the book probably summed up the general view of *The Twenty Years' Crisis* that it was as 'profound' as it was 'provocative' (*Times Literary Supplement*, 11 November 1939, p. 65).

In spite of, or perhaps even because of the various attacks on the book, it did remarkably well. The first edition of 1939 was reprinted three times in 1940, and a new (and controversially modified) edition came out in 1946. This in turn was reprinted on 12 occasions until 1978, before the reissue of the book in 1981. Carr thus had many good reasons – somewhere close to 50,000 in fact – to be pleased by a volume that not only sold well at home but also in the US. Indeed, Carr himself was well aware of the importance of the American market, and on at least two occasions in 1948 wrote to his publishers, Macmillan, urging them to keep the US market well stocked. This was in part because the book, he found out, was 'now being fairly widely used as a textbook', and also because of what he discovered at first hand to be the growth of a new discipline in American universities known as international relations! 'The study of the subject is increasing by leaps and bounds' and so 'the chances of sales are good', he noted. Hence, the 'more' copies of *The Twenty Years' Crisis* that Macmillan could 'supply America at present, the more are likely to be sold here'.

No doubt sales were much helped by an early assessment penned by one of the more senior figures in the new field of American international relations: Hans J. Morgenthau. Writing the first review article in the very first edition of the newly established journal *World Politics*, Morgenthau readily conceded the importance of the book and the sharp intellect of the man who had written it. Carr's contribution to political thought, he agreed, was 'of the first order'. Yet Carr in his view had failed to temper his own form of realism with a clear statement of moral purpose; consequently, he had become what Morgenthau called 'a Machiavelli without *virtu*', that is, someone who had no 'transcendent point of view from which to survey the political scene and to appraise the phenomenon of power'; see Hans J. Morgenthau, 'The Political Science of E. H. Carr', *World Politics*, Vol. 1, No. 1, October 1948, pp. 127–34. The influential Morgenthau piece was later reprinted in a shortened form, and with a different, rather more pungent title – 'The Surrender to the Immanence of Power: E. H. Carr' – appearing in his edited collection *Dilemmas of Politics* (Chicago: University of Chicago Press, 1962), pp. 350–7. For a later criticism of Carr, which takes up the issue of power and morality (and a good deal more), it is also useful to look at the quite lengthy essay by the American conservative academic Whittle Johnson, 'E. H. Carr's Theory of International Relations: A Critique', *Journal of Politics*, Vol. 29, 1967, pp. 861–84.

Carr, however, had more friends than he did enemies in the US, and one of the more enthusiastic was Kenneth Thompson, a prolific writer and editor, and a great admirer of the classical tradition of realism, the finest proponents of which in his view included Carr, Reinhold Niebuhr, Hans J. Morgenthau, Walter Lippmann and George F. Kennan. For a brief guide see in particular Kenneth Thompson, *Political Realism and the Crisis of World Politics* (New Jersey: Princeton University Press, 1960), and his edited collection *Masters of International Thought: Major Twentieth Century Theorists and the World Crisis* (Baton Rouge: Louisiana State University Press, 1980). Carr's other great American admirer was of course Robert

Gilpin, whose name was to become closely associated with the theory of hegemonic stability in the 1970s. Significantly, Gilpin derived this highly influential notion from his reading of Carr's *The Twenty Years' Crisis*, and then adapted it to international political economy. As he later admitted:

My interest in the relationship between the structure of the international political system and the nature of the international economy was first aroused by my reading of E. H. Carr, *The Twenty Years' Crisis*. In this classic study of the collapse of the open world economy at the outbreak of World War I and the subsequent inability of a weakened Great Britain to re-create a liberal international economy after the war, Carr demonstrated that a liberal world economy must rest on a dominant liberal power.

(Cited in Robert Gilpin, *Global Political Economy: Understanding the International Economic Order* (Princeton: Princeton University Press, 2001), p. 100)

In this context see Robert Gilpin, *War and Change in World Politics* (New York: Cambridge University Press, 1981) and *The Political Economy of International Relations* (Princeton: Princeton University Press, 1987).

Given his canonical status in the US as one of the high priests of classical realism, it is not surprising that a good deal was to be written about Carr's contribution to international relations theory. Useful starting points are: William C. Olsen, 'The Growth of the Discipline', in Brian Porter (ed.), *The Aberystwyth Papers, 1919–1969* (London: Oxford University Press, 1972), pp. 3–29; Roger Morgan, 'E. H. Carr and the Study of International Relations', in C. Abramsky (ed.), *Essays in Honour of E. H. Carr* (London: Macmillan Press – now Palgrave, 1974), pp. 171–82; and W. T. R. Fox, 'E. H. Carr and Political Realism', *Review of International Studies*, Vol. 11, 1985, pp. 1–16. Fox is one of the many to have commented negatively on what he thinks are the important differences between the 1939 and 1946 editions of *The Twenty Years' Crisis*. The critical chapter in Michael Joseph Smith's book, 'E. H. Carr: Realism as Relativism', in his *Realist Thought from Weber to Kissinger* (Baton Rouge: Louisiana State University Press, 1986), pp. 68–98, should also be read. Smith was one of the few Americans to refer to what he saw as some of the Marxist-influenced aspects of Carr's thought. However, for a more favourable analysis see J. D. B. Miller, 'E. H. Carr: The Realist's Realist', *The National Interest*, No. 35, Fall 1991, pp. 65–71. The title tells its own story about Miller's reading of Carr. R.G. Kauffman's 'E. H. Carr, Winston

Churchill, Reinhold Niebuhr and Us', *Security Studies*, Vol. 5, 1995, pp. 322–33 is also useful.

Ironically, while Carr was taken very seriously in the US, in the UK at least there never appeared to be the same level of engagement. No doubt his apparent indifference to the subject of international relations itself, and his known sympathy for the Russian revolution, made him something of a marginal figure in Cold War Britain. It was not insignificant, of course, that he was not invited to attend the first meeting establishing the British Committee for the Study of International Theory in 1960. Nor was it insignificant that one of the first really serious attempts to deal with *The Twenty Years' Crisis* was by someone who did attend: Hedley Bull. Furthermore, in a powerful critique originally entitled 'E. H. Carr and the Fifty Years' Crisis', but finally published in 1969 under the title 'The *Twenty Years' Crisis* Thirty Years On' (*International Journal*, Vol. 24, 1969, pp. 626–38), Bull called upon students of international relations to go beyond Carr and explore the idea that Carr rejected – 'the idea of international society itself' – and to use it as the starting point for a 'new analysis' of international relations. In the light of later discussions about Carr's intellectual relationship to the cause of the so-called 'English School', this review takes on added significance, as do the various criticisms directed at Carr's inability 'to keep the balance, the fruitful tension between power and morality' by that other high priest of the 'school', Martin Wight, in his *International Theory: the Three Traditions* (Leicester: Leicester University Press, 1991), pp. x, 16–18 and 267. In the same book Wight called *The Twenty Years' Crisis* 'the reflection of a diseased situation in Britain in the 1930s' (p. 267). The vexed issue of Carr's relationship, if any, with the 'English School' and the concept of an 'international society' can be followed up in Tim Dunne's important *Inventing International Society: a History of the English School* (London: Macmillan Press – now Palgrave, 1998). See the criticisms of Dunne's attempts to associate Carr with the 'school', and disassociate Manning, by Tony Brems Knudsen, 'Theory of Society or Society of Theorists? With Tim Dunne in the English School', *Cooperation and Conflict*, Vol. 35, No. 2, 2000, pp. 193–203; Samuel M. Makinda, 'International Society and Eclecticism in International Relations Theory', ibid., pp. 205–16; and Hidemi Suganami, 'A New Narrative, A New Subject? Tim Dunne on the "English School"', ibid., pp. 217–26. See also Tim Dunne's reply to critics in his 'Watching the Wheels Go Round: Replying to the Replies', *Cooperation and Conflict*, Vol. 36, No. 3, 2001, pp. 314–18.

Carr, it is true, never bothered to reply to either Bull or Wight. Nonetheless, we do know from his unpublished correspondence that he had a rather low opinion of the notion of an 'international society'. As he remarked in a letter to Stanley Hoffman in 1977, 'no international society exists, but an open club without substantive rules'. A few years earlier he had said much the same thing in an attack on C. A. W. Manning's book, later published under the title of *The Nature of International Society* (1962; London: Macmillan Press – now Palgrave, 1975).

'My fundamental criticism' [of his thesis about the existence of an international society], Carr wrote to the publishers even before they had decided to bring the volume out, 'is that the international game implied by it is in fact a creation of a small and influential group in the Western world during the past 400 years and there is no reason to suppose that it will survive the decline of these groups.

On Manning's contribution to, and role in, international relations in Britain, see the sympathetic account provided by Hidemi Suganami, 'C. A. W. Manning and the Study of International Politics', *Review of International Studies*, Vol. 27, No. 1, January 2001, pp. 91–108.

If Carr remained largely uninterested in international relations in Britain, not all practitioners of international relations were to remain so indifferent to him. This expressed itself from the 1970s onwards, however, not in a reaffirmation of his realism, but in a discovery of what many now said had for too long been ignored or misunderstood – namely, his critical attitude to power and those who exercised it. The first to hint at the puzzling and progressive character of Carr's realism was probably Graham Evans in his original but unfortunately largely ignored, 'E. H. Carr and International Relations', *British Journal of International Studies* (after 1980 *The Review of International Studies*), Vol. 1, No. 2, 1975, pp. 77–97. The critical and historical theme was then taken up by the radical political economist Robert Cox, a very early fan of Carr, in his influential manifesto-like article 'Social Forces, States, and World Orders: beyond International Relations Theory', *Millennium*, Vol. 10, No. 2, Summer 1981, pp. 126–55. Of equal significance was Ken Booth's inaugural lecture delivered in Aberystwyth in 1991, 'Security in Anarchy: Utopian Realism in Theory and Practice', later published in a slightly modified form in *International Affairs*, Vol. 67, No. 3, 1991, pp. 527–45. Andrew Linklater also played an important part in drawing attention to the more critical side of Carr. Among other things see his *Beyond Realism and Marxism: Critical Theory and International Relations* (London: Macmillan Press – now Palgrave, 1990); 'The Question of the Next Stage in International Relations Theory: A Critical-Theoretical Point of View', *Millennium*,

Vol. 21, 1992; and 'The Transformation of Political Community: E. H. Carr, Critical Theory and International Relations', *Review of International Studies*, Vol. 23, 1997, pp. 321–8. See also: Paul Howe, 'The Utopian Realism of E. H. Carr', *Review of International Studies*, Vol. 20, 1994, pp. 277–97; Richard Falk, 'The Critical Realist Tradition and the Demystification of Inter-State Power: E. H. Carr, Hedley Bull, Robert Cox', in Stephen Gill and James H. Mittelman, *Innovation and Transformation in International Studies* (Cambridge: Cambridge University Press, 1997); and Vendulka Kubalkova, 'The Twenty Years' Catharsis: E. H. Carr and IR', in Vendulka Kubalkova, Nicholas Onuf and Paul Kowert (eds), *International Relations in a Constructed World* (Armonk: Myron Sharpe, 1998), pp. 25–57. For an interesting, but brief discussion of Carr by a Marxist, see Justin Rosenberg's observations on Carr as one of many troubled realists in his *The Empire of Civil Society* (London: Verso, 1994), pp. 10–15.

Carr's ideas clearly continue to excite interest. His views on nationalism, for example, are discussed by Ernest Gellner in 'Nationalism reconsidered and E. H. Carr', *Review of International Studies*, Vol. 18, 1992, pp. 285–93, a sympathetic but not uncritical analysis. For a constructivist reading of Carr see Cecilia Lynch, 'E. H. Carr, International Relations Theory and the Societal Origins of International Legal Norms', *Millennium*, Vol. 23, No. 3, 1994, pp. 589–620, who attacks Carr for too easily dismissing the applicability of law, ethics and international organisations for the regulation of relations between states. It is also important to consult the three essays by Charles Jones, 'E. H. Carr through Cold War Lenses: Nationalism, Large States and the Shaping of Opinion', in Andrea Bosco and Cornelia Navari (eds), *Chatham House and British Foreign Policy 1919–1945: the Royal Institute of International Affairs During the Inter-War Period* (London: Lothian Foundation, 1994), pp. 163–85; 'E. H. Carr, Ambivalent Realist', in Francis A. Beer and Robert Hariman (eds), *Post-Realism: the Rhetorical Turn in International Relations* (East Lansing: Michigan State University Press, 1996), pp. 95–119; and 'Carr, Mannheim and a Post-Positivist Science in International Relations', *Political Studies*, Vol. XLV, 1997, pp. 232–46 where Jones explores the relationship between Carr and modern post-positivism through a reading of Mannheim. A provocative analysis of Carr's relationship to inter-war 'idealism' is also provided by the American scholar Brian C. Schmidt, *The Political Discourse of Anarchy: a Disciplinary History of International Relations* (New York: State University of New York Press, 1998). For a more general overview see Michael Cox, 'Will the real E. H. Carr please stand up?', *International Affairs*, Vol. 75, No. 3, July 1999, pp. 643–53. On realism – again – look at Randall L. Schweller and William C. Wohlforth,

'Evaluating Realism in Response to the End of the Cold War', *Security Studies*, Vol. 9, No. 3, Spring 2000, pp. 86–107 where the two Americans use Carr's thesis about 'peaceful change as an adjustment to the changed power relations' as a way of explaining the end of the Cold War. Finally, Stefano Guzzini's 'The Different Worlds of Realism in International Relations', *Millennium*, Vol. 30, No. 1, 2001, pp. 1–11, and Peter Wilson's 'Radicalism for a Conservative Purpose: The Peculiar Realism of E. H. Carr', ibid., pp. 123–36, provide a much wider perspective on realism in general and Carr in particular.

From the First to the Second Editions of The Twenty Years' Crisis: A Case of Self-censorship?

By the end of the Second World War, E. H. Carr had achieved something of a minor celebrity status. The first edition of *The Twenty Years' Crisis* had been reviewed widely and within a few months of publication in late 1939 had been printed four times, with early sales somewhere close to 8,000. By 1945 the book had even been translated into Spanish and Swedish. In 1942 Carr had then brought out his more forward-looking and altogether more popular study *The Conditions of Peace* – a powerful analysis of the world crisis with a set of policy recommendations to follow. Indeed, Carr was so confident of the message it contained that he requested that his publishers send a copy to President Franklin Roosevelt at the White House;[1] they in turn wondered whether or not he would be prepared to write a 'sequel', which was bound to be 'very popular both here and in the America'.[2] Carr declined. Finally, there were his regular leader columns in *The Times* of London, which by the end of the war had earned him the wrath of the Prime Minister, and the unofficial title of the 'Red Professor of Printing House Square' from those who neither liked his argument for economic planning, nor his ongoing advocacy of a post-war alliance between the Soviet Union, America and Britain. Carr may not have been a household name exactly, but he did have what one critic termed 'a strong position with the public'.[3] Certainly, it was strong enough to provoke an outburst by F. A. Hayek in his famous study *The Road To Serfdom*, published the following year. Carr, Hayek believed, was one of the 'totalitarians in our midst' whose 'contempt' for 'all ideas of liberal economists' was 'as profound as that of any' expressed by any German thinker in the past. In fact, according to Hayek, he was probably one of the more dangerous of that particular breed who were currently threatening the cause of political liberty and economic freedom in Britain;

[1] E. H. Carr to Macmillan Publishers, 16 March 1942, Macmillan Publishers Archives (hereafter MPA).
[2] Macmillan Publishers to E. H. Carr, 16 August and 23 August 1943 (MPA).
[3] Gilbert Murray to Lord David Davies, 31 May 1943. University of Wales Aberystwyth Archives.

consequently his views had to be taken very seriously indeed by all those who wished to defend both.[4]

Ever keen to make the most out of their very prolific rising star (in 1945 Carr had just signed a contract to write a study of Soviet Russia, in the same year he had brought out *Nationalism and After*, 1946 saw the appearance of another edition of his hugely popular *International Relations Between the Two World Wars*, as well as his *Soviet Impact on the Western World*), it made perfect sense for Macmillan to publish a new edition of the already well known *The Twenty Years' Crisis*. Carr was obviously enthusiastic too, and though not insensitive to the controversy the book had already created, was keen to see the volume back in the bookshops. On 23 July 1945, he thus returned the corrected galleys to Lovatt Dickson at Macmillan, and four months later, in November, sent back the corrected proofs with a new index made necessary by virtue of the fact that he had made some alterations to the original text of 1939.[5] However, as he explained in the preface to the second edition, these were of no great import. He wrote 'I have changed nothing of substance and have not sought to modify expressions of opinion merely on the grounds that I should not unreservedly endorse them today'. Yet he did concede that he had changed a few things. Nothing of significance perhaps. However, he had, to use his own words, 'recast' some 'phrases which would be misleading or difficult to readers now far remote in time from the original context', modified 'a few sentences which have invited misunderstanding', and removed 'two or three passages relating to current controversies which have been eclipsed or put in a different perspective by the lapse of time'.[6]

What were these modifications, and how substantial were they? Carr thought they were minor. Others since have disagreed, including the well known American academic William T. R. Fox, famous for many things, but especially for having coined the term 'superpowers' back in 1944. In the first E. H. Carr Memorial Lecture given in Aberystwyth in 1984, Fox argued that the deletions that Carr said were trivial were perhaps not trivial at all, but 'crucial'. Fox admitted to being 'shocked' when he discovered what had been omitted from the original edition of 1939. For what he discovered, and what made him indignant 'initially', was a series of cuts pertaining to the Munich agreement of 1938 and Carr's less than sensitive references to the fate of Czechoslovakia. Fox, to be fair, did

---

[4] F. A. Hayek, *The Road to Serfdom* (London: George Routledge and Sons, 1944), pp. 138–41.

[5] E. H. Carr to Macmillan Publishers, 23 July and 19 November 1945 (MPA).

[6] See 'Preface to the Second Edition', 15 November 1945, in this volume.

understand why Carr had made the changes. After all, *The Twenty Years' Crisis*, he reasoned, was a text of its time, which saw itself as an intervention into the policy debates of the late 1930s – on the side of appeasement – but which also contained what Carr hoped, and Fox agreed, were some fundamental statements about the nature of international relations. So deleting a few phrases here and there was perhaps no bad thing as this helped 'shed' the book's 'period piece aspect' and allowed it to 'stand on its more enduring merits'. The real issue in fact was not whether Carr should have made the changes in 1945 for the 1946 second edition, but whether he might have done so back in late 1939, just before the book was about to be published, but when it must have been perfectly obvious that appeasement as a policy was in tatters. However, as Fox went on to speculate, the reason Carr did not do so was possibly because he still hoped that the final plunge into total war could be avoided after the initial 'phony' skirmishes in late 1939. Thus there was no reason to delete the offending passages. By leaving them in he might even be able help avert the 'pointless carnage' that was bound to follow if his argument was either ignored by policy-makers or rendered irrelevant by Hitler's bombers.[7]

Fox was neither the first nor was he to be the last to comment on Carr's revisions. He also included a couple of the deleted passages by way of illustration. What he did not do – and in the context of a short article could not do – was to include all the deletions. Hence, while many students might be aware of the fact that certain changes were made, they are unable to judge how important these are, and whether or not they represent a major or minor modification to the original text of 1939. In the normal run of things this would not matter; except that in this particular case, the very existence of the alterations has led to suspicion regarding Carr and possibly even about the probity of the second edition of *The Twenty Years' Crisis* itself.

To clear up the 'mystery' of the missing passages, it might be useful to begin with a brief discussion of Carr's views on appeasement. As is well known, Carr had always felt that the inter-war system was deeply flawed, and that one of the reasons for this was Germany's impossible position as a 'have-not' state within it. Change to this system was therefore inevitable, and it was better that this occurred through negotiation around an agreed agenda rather than by any other means. This of course is why Carr was preoccupied with the whole notion of 'peaceful change' (the title of the crucial Chapter 13), and why he had been researching the

[7] W. T. R. Fox, 'E. H. Carr and political realism: vision and revision', *Review of International Studies*, Vol. 11, 1985, pp. 1–16.

subject so assiduously since he had resigned from the Foreign Office in early 1936. Indeed, in the same year that he had taken up his new academic post in Aberystwyth in Wales, he had become closely involved in a discussion group at Chatham House in London, the very purpose of which was to explore the issue.[8]

Thereafter, as his diaries reveal, he read widely on the subject in preparation for the volume that was finally published three years later in the form of *The Twenty Years' Crisis*. One could even argue that one of the many reasons why he wrote the book in the first place was to provide a theoretical, as opposed to a political explanation as to why peaceful change in the European system was a perfectly acceptable principle, and · that there was no reason why it should not be applied to Germany as well. Germany after all had been dealt a very tough hand at Versailles. The whole structure of the inter-war system had then been rendered non-viable by what Carr, among others, regarded as the indiscriminate use of the slogan of self-determination in 1919 (this is why he was especially critical of the Americans in general and Woodrow Wilson in particular). Moreover, it seemed to Carr, as it did to nearly everybody else within the political class at the time, that Britain's real interests in the world lay outside of Europe and not within it. Thus Britain had no serious reason for opposing Germany's legitimate desire to change a status quo that Carr never believed was defensible anyway. All this, and much more, about the real-world situation, including some fairly negative statements regarding the rights of small nations and the importance of big ones, can be found running like red threads through the whole of *The Twenty Years' Crisis*. Furthermore, these can be found in both editions of the book, which in terms of overall argument and structure are virtually identical.[9]

The changes that Carr made, therefore, need to be put into perspective. In no way do they alter the book's central claims. Nor do they do anything to modify his analysis of peaceful change in the international system. On the other hand, he does leave out some of the details and examples that he employed in the first edition; the reason, one must suspect, is because to those reading the volume in 1946 as opposed to 1939, they were bound to

---

[8] 'Discussion at Chatham House of Manning's views on peaceful change', 19 May 1936. Entry into the Carr Diaries, The University of Birmingham. The book that finally emerged from these discussions was edited by C. A. W. Manning under the title of *Peaceful Change* and was cited extensively by Carr in *The Twenty Years' Crisis*. Manning at the time was Montagu Burton Professor of International Relations at the London School of Economics. Carr later thanked Manning for reading and commenting on *The Twenty Years' Crisis*.

[9] One interesting little change that Carr did make was to change the designation of Hitler from 'Herr Hitler' in the 1939 edition to just 'Hitler' in the 1946 edition.

rankle somewhat. In fact, to be told that Munich was no bad thing after the searing experience of war and the failure of appeasement as a policy, might have annoyed a very large number of people indeed! Carr also made more than just one change to the original text, and while some of these were of no real importance at all, one or two were fairly blunt. Thus Carr decided to get rid of them, not to hide his opinions so much as to ensure that some of the more embarrassing sentences did not get in the way of the main thesis.

The bulk of the amendments are to be found in the chapter on peaceful change. Having discussed the issue in general terms, Carr then cites two examples of 'peaceful change' in action (in the first edition though not in the second).[10] The less controversial of the two (which remained in the second edition) was the nineteenth-century case of Bulgaria, which in 1877 had been deprived of much of the territory it had originally gained as a result of the Treaty of San Stefano. Whether or not this was fair did not much concern Carr: what did was that this change was brought about not by a war between the major powers of the time, but following 'discussions round a table in Berlin'. It was at this point that he made his first significant deletion by removing from the second edition a more modern example of another case where peaceful change had come about through negotiations around another table in Germany. He wrote in the first edition, but not in the second:

> If the power relations of Europe in 1938 made it inevitable that Czecho-Slovakia should lose part of her territory, and eventually her independence, it was preferable (quite apart from any question of justice or injustice) that this should come about as the result of discussions round a table in Munich rather than as the result either of a war between the Great Powers or of a local war between Germany and Czecho-Slovakia.

Carr could not be more explicit. Part of Czechoslovakia was going to be taken by Germany anyway, given the balance of power in Europe in 1938. Far better that it happen through reasoned debate between the 'great powers' rather than war. Thus in defence of the proposition that peaceful change was bound to be better than war – this being the ultimate purpose of statecraft after all – Carr felt no qualms about justifying Munich. Four paragraphs on, he then made yet another alteration in the context of a much wider discussion of what he called in both editions 'the failure to

[10] Compare pp. 278 (first edition, 1939), 219 (second edition, 1946) and 199 (2001 edition).

achieve a peaceful settlement with Germany in the period between the two world wars'. There was a broad consensus, he argued, that some parts of the settlement in 1919 had been 'just' and some 'unjust', and it was only a matter of time before the 'unjust' parts were rectified. 'Unfortunately' Germany did not have enough power in the 1920s to do anything to change the situation, and by the time it did, after Hitler assumed office, it addressed the problem in a rather different fashion. At this point Carr changed the wording somewhat, and in the second edition of 1946 he argued: 'By the time Germany had regained her power, she had adopted a completely cynical attitude about the role of morality in international politics.' In the original, first edition of 1939 he wrote:

> By the time Germany regained her power she had become – not without reason – almost wholly disillusioned about the role of morality in international politics.

He then added:

> There was not, even as late as 1936, any reasonable prospect of obtaining major modifications of the Versailles Peace Treaty by peaceful negotiation unsupported by the ultimatum or the fait accompli.

Here the main argument is not affected, but whereas in the new edition Carr merely describes Germany adopting a cynical attitude towards morality, in the first he almost appears to justify what Germany did. After all, her disillusionment with morality was, according to Carr in 1939 (but not in 1946), 'not without reason'. Moreover, whereas he seemed keen in every other circumstance to make the case for peaceful change, here he comes rather close to doing the opposite: that is, of providing a historical rationale of Germany threatening to use force. Indeed, if she had not acted in a threatening manner, says Carr, 'she' (Germany) might not have achieved what he already thought the status quo powers ought to have conceded through peaceful negotiations anyway. The change in words might be subtle, but the change in meaning is less so.[11]

This brings us to the third modification. In both editions Carr is critical of the way in which the status quo powers dealt with Germany in the late 1930s. According to Carr, it was not wrong for them to have acquiesced in what Germany did in Europe before 1938. What was wrong was to do so

---

[11] Compare pp. 281(first edition), 221 (second edition) and 201 (2001 edition).

under protest. This, he went on, 'inevitably created the impression that the remonstrating Powers acquiesced merely because they were unable or unwilling to make the effort to resist', not because they felt the changes were either 'reasonable and just'. Therefore what might have become a cause of 'reconciliation' became a factor in the 'further estrangement between Germany and the Versailles Powers', and so 'destroyed instead of increasing the limited stock of common feeling which had formerly existed'.

In the first edition, however, we also discover the following:

In March 1939, the Prime Minister admitted that in all the modifications of the Treaty down to and including the Munich Agreement, there was 'something to be said for the necessity of a change in the existing situation'. If, in 1935 and 1936, this 'something' had been clearly and decisively said, to the exclusion of the scoldings and protests by the official spokesmen of the status quo Powers, it might not have been too late to bring further changes within the framework of peaceful negotiation [and this, Carr continued] was a tragedy ... for which the sole responsibility cannot be laid at Germany's door.

Here Carr is not saying anything he had not said elsewhere before. However, whereas in 1946 he says nothing about direct responsibility, he does so in 1939 – and not all of it, he argues, is Germany's. Furthermore, by citing the Prime Minister in March 1939, Carr is also suggesting that whereas his predecessors had got things wrong (in another volume published at almost the same time as *The Twenty Years' Crisis*, Carr characterized their policy as oscillating 'between two opposite and incompatible extremes'),[12] the current British leader – Neville Chamberlain – might have got it right, or at least was prepared to accept what others had not: that Germany had a case and that measured change in the European order was both necessary and right.[13]

Carr then added another whole paragraph that did not appear in 1946. I quote in full:

The negotiations which led up to the Munich Agreement of September

---

[12] See E. H. Carr, *Britain: a Study of Foreign Policy from the Versailles Treaty to the Outbreak of War* (London: Longmans, Green and Co., 1939), p. 157. In the same volume Carr noted: 'Mr Chamberlain perceived more clearly than any other political leader the growing danger of a policy of words not matched either by willingness or a capacity to act' (p. 166)

[13] Compare pp. 281–2 (first edition), 222 (second edition) and 201 (2001 edition).

29, 1938, were the nearest approach in recent years to the settlement of a major international issue by a procedure of peaceful change. The element of power was present. The element of morality was also present in the form of the common recognition by the Powers, who effectively decided the issue, of a criterion applicable to the dispute: the principle of self-determination. The injustice of the incorporation in Czecho-Slovakia of three-and-a-quarter million protesting Germans had been attacked in the past by many British critics, including the Labour Party and Mr. Lloyd George. Nor had the promises made by M. Benes at the Peace Conference regarding their treatment been fully carried out. The change in itself was one that corresponded both to a change in the European equilibrium of forces and to accepted canons of international morality. Other aspects of it were, however, less reassuring. Herr Hitler himself seemed morbidly eager to emphasise the element of force and to minimise that of peaceful negotiation – a trait psychologically understandable as a product of the methods employed by the Allies at Versailles, but nonetheless inimical to the establishment of a procedure of peaceful change. The principle of self-determination, once accepted, was applied with a ruthlessness that left to Germany the benefit of every doubt and paid a minimum of attention to every Czecho-Slovak susceptibility. There was a complete lack of any German readiness to make the smallest sacrifice for the sake of conciliation. The agreement was violently attacked by a section of British opinion. Recriminations ensued on the German side; and very soon any prospect that the Munich settlement might inaugurate a happier period of international relations in which peaceful change by negotiation would become an effective factor seemed to have disappeared.

Here at least Carr makes some attempt to scold Hitler for having acted without due regard to Czechoslovak sensibilities after he had invaded part of the country in 1938. Indeed, he blames Hitler for not having used the opportunity presented by Munich to build a new, improved relationship with the other great powers. However, Hitler's actions are partly justified on the 'understandable' grounds that the Allies themselves had employed dubious methods at Versailles. Moreover, there was nothing wrong with what Germany then did in 1938. In fact, according to Carr, it was an almost textbook example of peaceful change. He even suggests that Czechoslovakia may have been partly to blame for what happened by not having treated well its own German minorities before Hitler decided to move in. He then adds some minor insult to injury by seeking to justify

Munich in terms of a principle with which he had always had some fundamental problems – namely, self-determination. Ironic indeed if you happened to be Carr. Less ironic than tragic if you happened to be a Czechoslovakian.[14]

This leads us, finally, to discuss something that Carr did not change at all, though he clearly wanted to – and perhaps should have done. As he pointed out in the preface to the second edition of 1946, his ideas about the role and position of the nation-state in the international order had undergone considerable modification since the beginning of the war and the original appearance of *The Twenty Years' Crisis*. While Carr conceded in the first edition that the 'concept of sovereignty' was 'likely' to 'become more blurred' in the future,[15] he still accepted that the nation-state in some form or another would probably remain the basic unit of the international system. As he put it in 1939, 'the nation is now, more than ever before, the supreme unit round which centre human demands for equality and human ambitions for predominance'.[16] The final collapse of the inter-war system, the experience of war, and his belief by 1945 in the need for a totally restructured international economy, inclined him to a more radical conclusion: that the nation-state, both small and large, 'was obsolete or obsolescent and that no workable international organization' could 'be built on a membership of a multiplicity of nation-states'. However, as he explained, he decided not to incorporate this shift into the new edition, and instead advised readers to look at his more recent book on the subject, *Nationalism and After*, published in 1945. Here they would be able to explore his 'present views' rather than those he had once held.[17]

It is of some interest, of course, that those who have been rather keen to point to those few passages where Carr provides an effective apologia for Munich, have said nothing about his equally important decision not to alter the text in a direction that might have better reflected his views on the nation-state. On the other hand, Carr himself made no reference to at least two alterations he did in fact make, one less significant than the other.

The first can be found in the final chapter where Carr questioned whether or not the nation-state would survive as the unit of power.

---

[14] Compare pp. 282–3 (first edition), 222 (second edition) and 201 (2001 edition).

[15] Cited in first edition on pp. 295–6; second edition, p. 230; 2001 edition, p. 212.

[16] First edition 1939, p. 291. In the second edition the 'now' was dropped. Carr thus wrote in 1946: 'The nation became, more than ever before, the supreme unit round which centre human demands for equality and human demands for predominance' (p. 227). 2001 edition, p. 210.

[17] See the 'Preface to the Second Edition', p. viii.

Following a lengthy discourse about the rise of the nation-state in history and its possible transcendence – 'few things are permanent in history and it would be rash to assume that the territorial unit is one of them' – he went on to discuss those factors that were undermining the nation-state and leading to 'the formation of ever larger political and economic units'. This tendency, he noted, 'appears to have been closely connected with the growth of large-scale capitalism and industrialism, as well as with the improvement of means of communication and of the technical instruments of power'. He then made a brief qualifying statement in 1939 (but not in 1946) about the complex character of nationalism within the larger context of the trend towards ever larger units whose

> course was marked by a conflict between the two functions of nationalism – the integrating and the disintegrating function. Prior to the middle of the nineteenth century, nationalism had been in the main a disintegrating force, breaking off fragment after fragment from the theoretical unit of mediaeval Christendom and making them into independent national units. Then nationalism, almost suddenly, passed over to an integrating role in Germany and Italy, and developed, in the most powerful countries, into an imperialism which seemed likely to divide the world into half dozen units of power.[18]

Why Carr should have dropped this is something of a mystery, unless he felt it was being a little too positive about nationalism as a factor in the making of the modern world. It is difficult to know, but does at least reveal the tensions in his own thought about the phenomenon. He then added (in both editions) that: 'The War threw this development into conspicuous relief'.

The second change came a couple of pages later in the same discussion. At this point Carr shifted the discussion away from an analysis of those various pressures that were leading to the 'concentration of political and economic power'. 'That is' only 'one side of the picture', he noted in the first edition of 1939; we also have to look, he said, 'in the other direction' where 'disintegrating forces may still be found at work'. He continued:

> It is a moot point whether the British Empire and the British Commonwealth of Nations are at present tending towards a strengthening or a relaxation of the bonds between the component parts. The United States are striving to extend their economic power by breaking

---

[18] See pp. 293–4 (first edition), 229 (second edition), 211 (2001 edition).

down foreign tariff barriers. But simultaneously, within the Union, new trade barriers – still insignificant, but ever increasing – are being surreptitiously raised between the states themselves. It is not certain that Germany will be able to constitute Central and South-Eastern Europe, or Japan Eastern Asia, into a compact economic unit, or that Soviet Russia will be able to knit together her vast territories as industrial development progresses.[19]

One of the main reasons why Carr may have decided to remove this passage was because in the intervening years between the publication of *The Twenty Years' Crisis* in 1939 and the new edition of 1946, certain changes had taken place: for example, Soviet Russia had been able to 'knit' its territories together, and the US after the war had not only begun to break down those external tariff barriers but had become more united itself. But it is also possible that he was keen to get rid of these few sentences here for another reason, for although they did not constitute a defence of the new empires then being created by Germany or Japan, nor were they especially critical. Admittedly, Carr did go on to suggest that they (and other imperial conglomerations) might not last because there 'may be a size which cannot be exceeded without a recrudescence of the disintegrating tendencies'.[20] However, this was not something that Carr seemed to look forward to with any enthusiasm. If anything, the logic of his argument led him to another, very different conclusion: that big units were good and small ones were bad – even if the units in question happened to be dominated by Nazi Germany and Imperial Japan. To have said as much in 1939 was problematic enough: to have repeated it in 1946 after the terrible price paid to eliminate both powers might have raised more than just a few eyebrows.

---

[19] See pp. 295 (first edition), 230 (second edition), 212 (2001 edition).
[20] First edition, p. 295; second edition, p. 230; 2001 edition, p. 212.

Glossary of Names

Sir Norman Angell (1873–1967)
British economist and worker for international peace who was awarded the Nobel Peace Prize in 1933. Angell is best known for his book *The Great Illusion* (1909), which marked a turning point for peace theory by seeking to refute the argument that conquest and war brought a nation great economic advantage. Angell attempted to show the opposite – that war could never profit anyone. In *The Twenty Years' Crisis* Carr attacked Angell for believing that war arose as a result of 'a lack of understanding'. Angell in turn thought Carr's book to be 'completely mischievous ... a piece of sophisticated moral nihilism', and later went on to strongly criticize it in his *Why Freedom Matters*, published in 1940. In this he referred to Carr, amongst other things, as being one of 'Hitler's intellectual allies in Britain'.

Arthur Balfour (1848–1940)
British Conservative prime minister from 1902–05. Balfour was brought into the coalition Cabinet as foreign secretary in 1916, and was a member of the British delegation at the Paris Peace Conference in 1919.

Edvard Benes (1884–1948)
Statesman, foreign minister, and founder of modern Czechoslovakia who forged its Western-oriented foreign policy between the First and Second World Wars but was forced to capitulate to Hitler's demands during the Czech crisis in 1938. As foreign minister he represented his country at the Paris Peace Conference in 1919, and championed the League of Nations throughout the inter-war period. He was to become president of Czechoslovakia in 1935.

Jeremy Bentham (1748–1832)
Economist, political and legal philosopher and social reformer, Bentham was the principal exponent of utilitarianism – a philosophy expressed in the idea that the best outcome in any society was that which provided the 'greatest good for the greatest number'. Bentham, according to Carr, took the eighteenth-century doctrine of reason and refashioned it to the needs of the nineteenth century. He also advocated a whole raft of social

reforms, many of which were later to be pursued by John Stuart Mill, one of Bentham's pupils. Carr saw Bentham as one of the founders of modern liberal 'utopianism', which found its early twentieth-century expression in Wilsonianism in the US.

William Jennings Bryan (1860–1925)

American politician and Woodrow Wilson's Secretary of State who firmly believed in arbitration to prevent war. An avowed pacifist, he eventually resigned after Wilson's second note to Germany in 1915 protesting about the sinking of the *Lusitania*.

N. M. Butler (1862–1947)

American educator, publicist and political figure who shared the Nobel Prize for Peace in 1931 and served as president of Columbia University from 1901 to 1945. Butler was a champion of international understanding, helping to establish the Carnegie Endowment for International Peace, of which he was a trustee and later president (1925–45). In 1930 he declared that the next generation would 'see a constantly increasing respect for Cobden's principles of free trade and peace', a view Carr strongly disparaged.

Edmund Burke (1729–97)

Best known for his critique of the revolution in France, *Reflections on the French Revolution*. Burke, despite his Whig sympathies and earlier support for the Americans in the War of Independence, has come to be seen as the principal exponent of Conservative philosophy.

Lord Cecil (1864–1958)

British statesman and winner of the Nobel Peace Prize in 1937, one of the principal draftsmen of the League of Nations Covenant in 1919, and one of the most loyal workers for the League until its supersession by the United Nations in 1945. Carr attacked him in *The Twenty Years' Crisis*, among other things, for his naïve faith in the League of Nations Covenant and belief in international conciliation.

Neville Chamberlain (1869–1940)

British Conservative prime minister (1937–1940) who was to be vilified by many for his policy of appeasement towards Nazi Germany. In particular he is remembered for the agreement he concluded with Hitler at Munich in 1938. He was harried by anti-appeasers on the Conservative benches. In September 1939 he was forced to declare war, but was ill-suited for the task ahead of him. A parliamentary revolt ensured his resignation in 1940. Six months later he died a broken man. Carr at the time was altogether more generous towards Chamberlain than later critics. Thus when Chamberlain became prime minister in May 1937, Carr wrote two years later in his *Britain: a Study of Foreign Policy from Versailles to the Outbreak of War* that Chamberlain was a 'realist' who 'perceived more clearly than any other political leader the growing danger of a policy of words not matched by deeds'. In *The Twenty Years' Crisis* Carr praised the Munich Agreement signed by Chamberlain and Hitler on 29 September 1938 as being 'the nearest approach in recent years to the settlement of a major international issue by the procedure of peaceful change'.

Winston Churchill (1874–1965)

Best known for guiding Britain to victory during the Second World War from 1940–45. However, Churchill had a tumultuous political career long before that. He served as a Liberal minister under Asquith and Lloyd George, and was First Lord of the Admiralty when the First World War broke out. He served in several different ministerial positions during the coalition governments, and went on to become Chancellor of the Exchequer from 1924–29. He was a backbencher throughout the 1930s, though he became an increasingly vociferous opponent of appeasement towards Nazi Germany. The outbreak of the Second World War led to his being returned to the Admiralty, to become prime minister in 1940. Strongly criticized by Carr during the war for having no programme for post-war economic reform, Churchill later attacked Carr and *The Times* (for which Carr worked during the Second World War) for being critical of British foreign policy in Greece. Carr in turn attacked Churchill's famous 'Iron Curtain' speech made at Fulton, Missouri, in the March of 1946.

Richard Cobden (1804–65)

British politician known for his successful fight for the repeal of the Corn Laws and his defence of free trade. With John Bright, he believed that free trade would result in the reduction of armaments and the promotion of international peace. Theoretically he had influence on the post-First World War 'utopians' such as Nicholas Murray Butler whom Carr criticised in *The Twenty Years' Crisis*.

Confucius (551–479 BC)

Chinese philosopher who emphasized the importance of moral values in all social and political order. He believed that the hierarchical structure of traditional Chinese society was natural, though he argued that membership of the ruling class should be moral rather than hereditary.

Peter Drucker (1909–)
American management consultant, educator and author whose later writings provided much of the philosophical and practical underpinning of the modern business corporation. In the 1930s, however, Drucker was far from the conservative he was later to become, and was admired by Carr for having written *The End of Economic Man*.

Anthony Eden (1897–1977)
Rapidly promoted through the Conservative ranks during the 1930s, he became foreign secretary in 1935. He resigned three years later in opposition to Chamberlain's policy of appeasement. Eden was brought back into the Cabinet at the outbreak of war in 1939 as dominions secretary, and once again became foreign secretary after Chamberlain's resignation in 1940. Eventually he became prime minister in 1955. Carr worked closely with Eden while he was still in the Foreign Office until his own resignation in early 1936.

Friedrich Engels (1820–95)
Mainly remembered as a life-long friend, collaborator and financial supporter of Karl Marx. Like Marx, Engels was critical of earlier Socialists such as Robert Owen, Saint-Simon and Fourier, and opposed them for their 'utopianism'. Carr, it seems, borrowed the notion and transplanted it into his own critique of those early writers on international relations who in his view were 'utopian' too.

T. H. Green (1836–82)
English political philosopher who, according to Carr, 'tempered the doctrines' of Hegel with British nineteenth-century liberalism. Carr also quoted him as one of those who accepted that the morality of individuals can only be social morality – and social morality implies duty to fellow members of all other communities, including the larger international community. But, as Carr asked of Green in *The Twenty Years' Crisis*, 'In what sense can we find a basis for international morality by positing a society of states?'

Hegel (1770–1831)
German philosopher who developed a dialectical scheme that emphasized the progress of history and ideas from thesis to antithesis, and thence to a

higher and richer synthesis. He was one of the great modern creators of a philosophical system that influenced the development of Marxism in particular. Carr had a high opinion of Hegel, whose work he had read in the early 1930s in preparation for his earlier books: *The Romantic Exiles* (1933), *Karl Marx* (1934) and *Michael Bakunin* (1937). Carr saw Hegel – along with Marx – as belonging to the 'historical school' of realists; that is, those who championed the view that no ethical standards are applicable to relations between states, an idea that certainly influenced him in writing *The Twenty Years' Crisis*.

Leonard Hobhouse (1864–1929)
Political philosopher and activist, a crucial figure in the emergence of 'social liberalism', which justified increased state intervention as necessary for the achievement of both social and individual goods, as well as social order. Hobhouse taught at the London School of Economics and had an influence on a number of Liberal reformers whom Carr criticized in *The Twenty Years' Crisis*.

Cordell Hull (1871–1955)
F. D. Roosevelt's secretary of state from 1934–44. Famous in the 1930s for his efforts to maintain some form of open trading system, he was quoted as once saying 'that when goods crossed frontiers, armies didn't'.

Immanuel Kant (1724–1804)
German philosopher of the idealist school. In terms of international relations, he argued for a 'league of nations' to enforce the natural, derivable and international law, envisaging a decline in the power of individual states as that of the universal authority came to be established. Kantianism is identified in modern international relations as standing in opposition to state-centric realism.

John Maynard Keynes (1883–1946)
Liberal British economist who first became famous for his *The Economic Consequences of the Peace* – a savage attack on the Versailles Peace Treaty of 1919 with which Carr fully agreed. Keynes went on in the 1930s to write his *General Theory*, which proposed state intervention to overcome large-scale unemployment.

Sir Hersch Lauterpacht (1870–1924)
British international lawyer of the Grotian school, criticized by Carr for his belief that the events of the 1930s were only a temporary regression of international order that would eventually be overcome.

Lenin (1870–1924)

Leader of the Bolshevik revolution for whom Carr came to have some regard – both as a state builder and a political theorist of revolution. Interestingly, Carr quotes Lenin over half a dozen times in *The Twenty Years' Crisis*, often agreeing with what he has to say. He quotes Lenin on law. Lenin insists that law does not reflect any fixed ethical standard but is rather 'the formulation, the registration of power relations, an expression of the will of the ruling class'.

Maxim Litvinov (1876–1951)

Soviet diplomat and commissar of foreign affairs (1930–39) who was a prominent advocate of world disarmament and of collective security with the Western powers against Nazi Germany before the Second World War.

David Lloyd George (1863–1945)

Born in Wales, Lloyd George became prime minister in the midst of war in 1916, and went on to lead Britain to victory. He subsequently headed the British delegation at the Paris Peace Conference, and continued as prime minster until 1922 when the Conservatives withdrew from the governing coalition.

John Locke (1632–1704)

English philosopher widely regarded as one of the fathers of the Enlightenment and as a key figure in the development of liberalism. Locke was mainly concerned with the proper extent of freedom and religious toleration. He believed that the political authority's fundamental role was to protect property. Locke defended the concept of private property on the grounds that it was legitimated when it was mixed with the individual's labour.

Niccoló Machiavelli (1469–1527)

Florentine political adviser and historian. In his best known works *The Prince* and the *Discourses*, he outlined a theory of *raison d'état* where he seemed to be suggesting that the use of any technique was permissible as long as it achieved the desired end. The term 'Machiavellian' later entered the world's language to designate cynicism, manipulation and duplicity. Carr was much influenced by Machiavelli and indeed used him as the basis for constructing his own realist critique of what he termed 'utopianism'. Carr described him as 'the first important political realist'.

Karl Mannheim (1893–1947)
Hungarian sociologist who made an important contribution to the sociology of knowledge. Like Marx, Mannheim wanted to relate belief systems that emerged in political historical periods to the socio-economic and political conditions that seemed to stimulate and sustain them. Unlike Marx, though, Mannheim believed utopianism to be a forward-looking and visionary tendency that was capable of breaking out of the constraints of the existing social order and could thus point to the possibility of real change in the historical process. The influence of Mannheim's *Ideology and Utopia* on Carr was immense. Not only did Carr want to call his own book *Utopia and Reality* rather than *The Twenty Years' Crisis* (this was his publisher's suggestion), he also used Mannheim's central theoretical claim about the relativity of knowledge as the basis of his own thinking about ideas.

Charles Manning
Montagu Burton Professor of International Relations at the London School of Economics between 1930 and 1962 – and a friend of Carr's with whom Carr worked quite closely in the 1930s when developing his own views on peaceful change. Manning read drafts of *The Twenty Years' Crisis*, though he was by no means in agreement with all of its central claims. Carr in turn did not share Manning's views about the existence of a larger 'international society of states'.

Alfred Marshall (1842–1924)
British economist, one of the chief founders of the school of English neoclassical economics. Marshall is often considered to have been in the line of descent of the great British economists – Adam Smith, David Ricardo and J. S. Mill.

Karl Marx (1818–83)
German philosopher, sociologist, socialist and economist. His influence right across the social sciences has been immense. Fundamentally, Marx was a historical materialist, believing that history could be studied as a scientific process. Marx argued that societies would progress through several stages of socio-economic development. Capitalism was the highest stage, though Marx believed that it contained its own contradictions that would lead it to implode, to be replaced by socialism. Carr was not a Marxist and disagreed with both Marx's labour theory of value and his idea of the proletariat as a revolutionary class. Nonetheless, Carr borrowed heavily from Marx's historical sociology and from his various insights

about the failure of capitalism as a system in the inter-war period. Two years before his death, in 1980, Carr confessed that *The Twenty Years' Crisis* 'was not exactly a Marxist work', but was 'strongly impregnated with Marxist ways of thinking applied to international affairs'. Carr often contrasted Marx the 'realist' with the anarchist Bakunin as 'utopian'.

James Mill (1773–1836)

Liberal thinker, mainly remembered now for the education of his son, John Stuart Mill, and his life-long friendship with Jeremy Bentham.

John Stuart Mill (1806–73)

Liberal philosopher and political activist, strongly influenced by the work of Bentham. However, Mill was to differ from Bentham in his belief in individual personal character and self-regarding personal conduct. Mill was also the leading Liberal feminist of his day.

Reinhold Niebuhr (1892–1971)

American theologian and political scientist, who in the 1930s at least employed critical Marxist categories to understand the modern world. Carr always remained a warm admirer of Niebuhr and was strongly influenced by his 1932 study *Moral Man and Immoral Society. The Twenty Years' Crisis* is littered with various hard-hitting quotes from Niebuhr about the hypocrisies of the privileged and the exercise of power. Later Niebuhr moved away from his earlier attachment to Marxism but he continued to have a huge influence in the US in the post-war period.

Robert Owen (1771–1858)

British social reformer whose ideas contributed to the development of nineteenth-century socialism. Owen was essentially a communitarian Socialist who believed that the amelioration of social conditions and intelligent organization of the labour process were the necessary means for the creation of greater equality and justice. Like Engels, Carr regarded him as a utopian Socialist whose contribution to nineteenth-century thought was not his realistic analysis of society but his desire and aspiration to change reality.

Plato (427–347 BC)

Greek philosopher whose most important treatise, the *Republic*, espoused Plato's view of the ideal society. Carr saw him as an early example of a type of utopian who proposed what he called in *The Twenty Years' Crisis* 'highly imaginative solutions whose relation to existing facts was one of flat negation'.

Jean-Jacques Rousseau (1712–78)

Moral, political and educational philosopher whose political ideas are exemplified in his *Social Contract*. His main concern was liberty, and he believed that it could be attained through 'general will'. This is the will of each individual in favour of the good of the whole community, and is superior to his own particular interests.

Sir John Simon (1873–1954)

Liberal politician who served as foreign secretary from 1931–35. He was Chancellor of the Exchequer in Neville Chamberlain's government from 1937–40, and closely identified with the policy of appeasement towards Nazi Germany.

Saint-Simon (1760–1825)

French social theorist and one of the chief founders of Christian socialism. In his major work, *Nouveau Christianisme* (1825), he proclaimed a brotherhood of man that must accompany the scientific organization of industry and society. Carr regarded Saint-Simon as another important example of a utopian socialist who aspired to change everything without first having provided a realistic analysis of the world.

Adam Smith (1723–90)

Scottish philosopher and founder of classical economics. In his work, Smith argued in his famous *The Wealth of Nations* that the individual's pursuit of private gain would, unintentionally through the 'invisible hand', benefit the whole of society. He also described the importance of the division of labour. Carr viewed Smith's particular brand of *laissez-faire* economics as being responsible for popularizing the doctrine of the 'harmony of interests'.

Jan Smuts (1870–1950)

Smuts fought against the British during the Boer War but subsequently became a firm supporter of the Union of South Africa that was established in 1910. He resisted calls for neutrality during the First World War and served in the British army. In 1917 he became a member of the British War Cabinet. Following the war he became prime minister until 1924, and was active in South African politics throughout the inter-war period, becoming prime minister again at the outbreak of the Second World War.

Sophocles (496–406 BC)
One of the three great tragic playwrights of classical Greece, ranking with Aeschylus and Euripides. He wrote some 123 dramas, only 7 of which have survived.

Stoics
The Stoics taught that material possessions were of no importance whatsoever for a man's happiness. The basis of human happiness, they believed, was to live 'in agreement' with oneself, a statement that was later replaced by the formula 'to live in agreement with nature'. The only real good for man is the possession of virtue; everything else (wealth or poverty, health or illness, life or death) is of little consequence.

William Taft (1857–1930)
Republican president of the US from 1909–13, Taft encouraged 'dollar diplomacy' abroad, relying on trade rather than military power to spread influence. However, he is mainly remembered for splitting the Republican Party when his former friend and predecessor, Teddy Roosevelt, decided to run against him in 1912 as a Progressive, splitting the Republican vote and allowing Woodrow Wilson to come to office.

Arnold Toynbee (1889–1975)
British historian, best known for his study of the rise and fall of civilizations over the course of history. He concluded that civilisations rose by responding successfully to challenges under the leadership of creative minorities composed of elite leaders. Civilizations declined when their leaders stopped responding creatively, and the civilizations declined and died. In the 1930s he exercised great influence through Chatham House – where he was director of studies from 1924–54 – and through its annual *Survey of International Affairs*, which he wrote single-handedly. A liberal internationalist and opponent of appeasement, Carr attacked him frequently in *The Twenty Years' Crisis*. Toynbee later argued that 'Carr is the consummate debunker who was debunked by the war itself'.

Woodrow Wilson (1856–1924)
Democratic president of the US from 1913–21. Mainly remembered for taking his country into the First World War in 1917. Wilson enunciated 'Fourteen Points' for achieving peace and led the American delegation at the Paris Peace Conference in 1919, at which he attempted to realize his vision of a new international order. While failing to achieve everything he wanted, his main objective of establishing a League of Nations did come

into being, though the Senate vetoed the possibility of American participation in the organization. Carr was a major critic of Wilson on two counts: first he opposed what he saw as Wilson's indiscriminate use of the slogan of the right of self-determination, which he regarded as economically problematic and politically destabilizing in the inter-war period; second, he saw Wilson as the quintessential moralist in politics who used fine-sounding slogans as a means of advancing American power in the world. On one level *The Twenty Years' Crisis* can be read as an extended attack on Wilson and Wilsonianism – this in spite of the fact that Carr held the Woodrow Wilson Chair in International Politics at Aberystwyth from 1936–47.

Sir Alfred Zimmern (1879–1957)
First holder of the Woodrow Wilson Chair at the University of Wales, Aberystwyth, in 1919. Zimmern was a classicist who wrote several books in the inter-war period on international affairs, his best known being *The League of Nations and the Rule of Law* (1936). Carr and Zimmern stood at opposite ends of the theoretical and political spectrum. Carr was a radical, increasingly critical of the League of Nations, who attacked the Liberal Zimmern in *The Twenty Years' Crisis* for failing to grasp the deeper cause of the inter-war crisis. Zimmern was a 'cautious idealist', who saw Carr as 'a thorough-going relativist with no moral compass to guide him'.

Chronology

1919

Jan. 8   Opening of the Paris Peace Conference under the Chairmanship of Georges Clemenceau.

June 28   German representatives sign the Peace Treaty in the Hall of Mirrors at the Palace of Versailles.

Nov. 19   US Senate votes against ratification of the Treaty of Versailles, thereby leaving the US outside the League of Nations.

1920

Jan. 10   Ratification of the Treaty of Versailles brings the League of Nations into existence, with 29 initial members (out of 32 Allied signatures to the Versailles Treaty).

Jan. 16   US Senate votes against joining the League of Nations.

Feb. 26   In accordance with the Treaty of Versailles, the League of Nations takes over the Saar area between France and Germany; France takes control of the Saar's coal deposits.

March 19   US Senate rejects the Versailles Treaty.

April 6–May 17   While German troops are suppressing a rebellion in the Ruhr (a demilitarized area), French troops occupy Frankfurt, Darmstadt, and Hanau until German forces have withdrawn.

April 25   Supreme Allied Council disposes territories formerly in the Ottoman Empire: it assigns mandates over Mesopotamia and Palestine to Britain, and over Syria and the Lebanon to France.

Nov. 15   Danzig (modern Gdansk) is proclaimed a free city.

1921

Jan. 6   End of Russian civil war, with victory to the Bolsheviks.

Jan. 24–29   Paris Conference fixes Germany's reparation payments.

March 8	French troops occupy Düsseldorf and other towns in the Ruhr because of Germany's failure to make preliminary reparations payment.
April 27	Reparations Commission fixes Germany's liability at 132,000 million gold Marks.
May 6	Peace treaty signed between Germany and Russia.
Sept. 30	French troops evacuate the Ruhr.
Nov. 12–Feb. 6 1922	Washington conference on disarmament.
Dec. 29	US, British Empire, France, Italy and Japan sign Washington Treaty to limit naval armaments.

1922

April 16	Rapallo Treaty between Germany and Russia: Germany recognizes Russia as 'a great power' and both sides waive reparations claims; the Treaty leads to the resumption of diplomatic and trade relations, and to co-operation between the two countries' armies.
Oct. 28	In Italy, Fascists march on Rome, leading to Mussolini forming a government composed of Liberals and Nationalists, as well as Fascists, at the King's invitation on 31 October.

1923

Jan. 11	As a result of Germany's failure to meet reparations payments, French and Belgian troops occupy the Ruhr; Germans respond with passive resistance and sabotage; the occupiers make arrests and deportations, and cut off the Rhineland from the rest of Germany.
Nov. 8–9	The 'Munich Putsch': Adolf Hitler and the National Socialists attempt a *coup d'état* in Munich.

1924

Jan. 21	Death of Soviet leader Lenin. Stalin takes over.
July 16	At London conference on reparations, the Dawes Plan, which removes reparations from the sphere of political controversy, is approved.
Aug. 16	French delegates at London conference agree to evacuate the Ruhr within a year.

1925

Oct. 5–16   Locarno Conference discusses the question of a security pact and strikes a balance between French and German interests by drafting treaties (a) guaranteeing the French–German and Belgian–German frontiers; (b) between Germany and France, Belgium, Czechoslovakia and Poland respectively; (c) regarding a mutual guarantee between France, Czechoslovakia and Poland. Britain is involved in the guarantee of Franco-Belgian-German frontiers but not in the arrangements in Eastern Europe.

Dec. 1   Locarno treaties signed in London.

1926

Sept. 8   Germany is admitted into the League of Nations.

1927   Stalin consolidates his power in Soviet Russia, and in China the first revolution is crushed.

1928   The Kellogg–Briand Pact outlawing war and providing for pacific settlement of disputes is signed in Paris by 65 states, including the US and the USSR.

1929

June 7   Young Committee reviewing German reparations payments recommends that Germany should pay annuities, secured on mortgage of German railways, to an international bank until 1988.

Aug. 6–13   At a Reparations Conference at the Hague, Germany accepts the Young Plan; Allies agree to evacuate the Rhineland by June 1930.

Oct. 24–29   Crashes in share values on Wall Street stock market, New York, starting with 'Black Thursday' and continuing (after closure of the market from noon on 24 October until 28 October) on 'Black Monday' (28 October) and 'Black Tuesday' (29 October). Leads to the cessation of American loans to Europe.

1931

Sept. 18   Japanese invasion of Manchuria.

1933

Jan. 30   Adolf Hitler is appointed Chancellor of Germany; his cabinet includes only two Nazis, Herman Goering (Minister without Portfolio) and Wilhelm Frick (Minister of the Interior). Franz von Papen is Vice-chancellor, Constantin von Neurath foreign minister.

March 4   Inauguration of F. D. Roosevelt; Cordell Hull is appointed Secretary of State.

March 27   Japan announces that it will leave the League of Nations (effective from 1935).

Oct. 14   Germany withdraws from the League of Nations and its Disarmament Conference.

1934

July 25   Engelbert Dollfuss, Chancellor of Austria, is murdered in an attempted Nazi *coup d'état.*

Sept. 18   USSR is admitted into the League of Nations.

1935

Jan. 13   Plebiscite in the Saarland: 90.8 per cent of voters favour incorporation into Germany.

March 16   Germany repudiates disarmament clauses of the Versailles Treaty and introduces conscription.

May 2   France–USSR treaty of mutual assistance for five years.

May 16   USSR–Czechoslovakia pact of mutual assistance.

July 25–Aug. 20   Meeting of the Third International (Soviet-controlled international Communist organization) declares that Communists in democratic countries should support their governments against Fascist states.

Oct. 2   Italy invades Ethiopia.

Oct. 7   Following Italy's invasion of Ethiopia, the League of Nations Council declares and denounces Italy as the aggressor.

Oct. 19   League of Nations imposes sanctions against Italy.

1936

Jan. 6–March 25   Resumption of London Naval Conference: Japan withdraws on 15 January because other countries refuse to accept its demand for a common upper limit on naval strength.

March 7	German troops occupy the demilitarized zone of the Rhineland, thereby violating the Treaty of Versailles.
July 17	Army mutiny in Spanish Morocco, led by Franco, to uphold religion and traditional values; other mutinies occur throughout Spain, thereby starting the Spanish Civil War.
Nov. 1	Following the visit of Italian Foreign Minister Ciano to Berlin, Italian Prime Minister Benito Mussolini proclaims the Rome–Berlin Axis.
Nov. 25	Germany and Japan sign Anti-Comintern Pact; the countries agree to work together against international communism; Germany also recognizes Japan's regime in Manchuria.

1937

May 28	On Stanley Baldwin's retirement, Neville Chamberlain forms a National Government, with Sir John Simon as Chancellor of the Exchequer and Anthony Eden as Foreign Secretary.
July 7	In China, incident at Marco Polo Bridge, southeast of Beijing, is followed by Japanese invasion of northeast China.
July 17	Naval agreements between Britain and Germany, and Britain and the USSR.
Nov. 6	Italy joins German–Japanese Anti-Comintern Pact.
Nov. 17–21	British cabinet minister Lord Halifax meets Adolf Hitler to attempt peaceful settlement of the Sudeten problem.
Dec. 11	Italy withdraws from the League of Nations.

1938

Feb. 20	British Foreign Secretary Anthony Eden resigns in protest at Prime Minister Neville Chamberlain's priorities in foreign affairs. Chamberlain had declined President Roosevelt's suggestion of a conference on international relations and was determined to obtain an agreement with Italy; Eden is succeeded on 25 February by Lord Halifax.
Feb. 21	Winston Churchill leads an outcry in the House of Commons against Chamberlain and the following day 25 members of the administration vote against the government in a censure motion.

March 12	German troops enter Austria, which the next day is declared part of the German Reich.
Sept. 15	British Prime Minister Neville Chamberlain visits Adolf Hitler at Berchtesgaden; Hitler states his determination to annex the Sudetenland in Czechoslovakia on the principle of self-determination.
Sept. 29–30	The Munich Conference, when Chamberlain, French Prime Minister Édouard Daladier, Adolf Hitler and Benito Mussolini agree to Germany's military occupation of the Sudetenland, while the remaining frontiers of Czechoslovakia are guaranteed; Germany becomes the dominant power in Europe and both the Little Entente and the French system of alliances in Eastern Europe are shattered; on his return to London, Chamberlain declares that he has bought 'peace with honour. I believe it is peace in our time'.

1939

March 15	German troops occupy Bohemia and Moravia in Czechoslovakia; Hitler makes a triumphal entry into Prague in the evening; the region becomes a protectorate ruled by Constantin von Neurath.
March 16	Slovakia is placed under German 'protection', while Hungary annexes Ruthenia (formerly part of Czechoslovakia).
March 21	Germany demands of Poland that Germany should acquire the Free City of Danzig and routes through the 'Polish corridor' (which provides Poland with access to the Baltic); Poland rejects the demands.
March 28	In the Spanish Civil War, Madrid surrenders to nationalists; remaining republican areas surrender the next day, ending the war.
March 28	Adolf Hitler denounces the non-aggression pact with Poland (of January 1934).
March 31	Britain and France pledge to support Poland in any attack on Polish independence (pact of mutual assistance agreed on 6 April).
April 18	USSR proposes triple alliance with Britain and France.
April 28	Hitler denounces 1935 British–German naval agreement and repeats demands on Poland.

May 22	Hitler and Mussolini sign ten-year military alliance (the 'Pact of Steel').
Aug. 23	Nazi–Soviet Pact; the parties agree not to fight each other; secret protocols provide for the partition of Poland and for the USSR to operate freely in the Baltic States, Finland, and Bessarabia.
Aug. 25	British–Polish treaty of mutual assistance signed in London.
Sept. 1	Germany invades Poland and annexes Danzig; Italy declares neutrality.
Sept. 3	Britain and France declare war on Germany, following Germany's failure to reply to ultimata on Poland; Australia and New Zealand also declare war.

TO THE MAKERS
OF THE COMING PEACE

Philosophers make imaginary laws for imaginary commonwealths, and their discourses are as the stars which give little light because they are so high.

BACON, *On the Advancement of Learning*

The roads to human power and to human knowledge lie close together and are nearly the same; nevertheless, on account of the pernicious and inveterate habit of dwelling on abstractions, it is safer to begin and raise the sciences from those foundations which have relation to practice, and let the active part be as the seal which prints and determines the contemplative counterpart.

Id., Novum Organum

Preface to the 1981 Printing

A generation later, I have little to add to the preface which I wrote for the second edition. In 1945 *The Twenty Years' Crisis* was already a period piece, and such it remains. But it was an attempt to navigate then relatively uncharted waters; and, though its observations were cast in the mould of a brief historical period (as all general observations tend to be), they perhaps retain a wider interest and significance. The reflexions on nationalism in the last paragraph of my 1945 preface have been qualified, but not altogether falsified, by the unforeseen emergence of mutually antagonistic superpowers.

E. H. CARR

1 *August* 1980

Preface to Second Edition

The demand for a second edition of *The Twenty Years' Crisis* faced the author with a difficult decision. A work on international politics completed in the summer of 1939, however rigorously it eschewed prophecy, necessarily bore marks of its time in substance, in phraseology, in its use of tenses and, above all, in such phrases as 'the War', 'post-War' and so forth, which can no longer be related without a strong effort on the part of the reader to the war of 1914–18. When, however, I approached the task of revision, it soon became clear that if I sought to rewrite every passage which had been in some way affected by the subsequent march of events, I should be producing not a second edition of an old book but essentially a new one; and this would have been a clumsy and unprofitable attempt to force new wine into old bottles. *The Twenty Years' Crisis* remains a study of the period between the two wars written as that period was coming to an end and must be treated on its merits as such. What I have done, therefore, is to recast phrases which would be misleading or difficult to readers now far remote in time from the original context, to modify a few sentences which have invited misunderstanding, and to remove two or three passages relating to current controversies which have been eclipsed or put in a different perspective by the lapse of time.

On the other hand, I have changed nothing of substance, and have not sought to modify expressions of opinion merely on the ground that I should not unreservedly endorse them to-day. Perhaps, therefore, I may be permitted to indicate here the two main respects in which I am conscious of having since departed to some degree from the outlook reflected in these pages.

In the first place, *The Twenty Years' Crisis* was written with the deliberate aim of counteracting the glaring and dangerous defect of nearly all thinking, both academic and popular, about international politics in English-speaking countries from 1919 to 1939 – the almost total neglect of the factor of power. To-day this defect, though it sometimes recurs when items of a future settlement are under discussion, has been to a considerable extent overcome; and some passages of *The Twenty Years' Crisis* state their argument with a rather one-sided emphasis which no longer seems as necessary or appropriate to-day as it did in 1939.

Secondly, the main body of the book too readily and too complacently accepts the existing nation-state, large or small, as the unit of international society, though the final chapter offers some reflexions, to which subsequent events have added point, on the size of the political and economic units of the future. The conclusion now seems to impose itself on any unbiased observer that the small independent nation-state is obsolete or obsolescent and that no workable international organisation can be built on a membership of a multiplicity of nation-states. My present views on this point have been worked out in a small book recently published under the title *Nationalism and After*, and I can therefore with the better conscience take the only practicable course and leave the present work substantially as it was completed in 1939.

E. H. CARR

15 *November* 1945

Preface to First Edition

This book, which was originally planned in 1937, was sent to the press in the middle of July 1939 and had reached page proof when war broke out on 3 September 1939. To introduce into the text a few verbal modifications hastily made in the light of that event would have served little purpose; and I have accordingly preferred to leave it exactly as it was written at a time when war was already casting its shadow on the world, but when all hope of averting it was not yet lost. Wherever, therefore, such phrases as 'the war', 'pre-war' or 'post-war' occur in the following pages, the reader will understand that the reference is to the War of 1914–18.

When the passions of war are aroused, it becomes almost fatally easy to attribute the catastrophe solely to the ambitions and the arrogance of a small group of men, and to seek no further explanation. Yet even while war is raging, there may be some practical importance in an attempt to analyse the underlying and significant, rather than the immediate and personal, causes of the disaster. If and when peace returns to the world, the lessons of the breakdown which has involved Europe in a second major war within twenty years and two months of the Versailles Treaty will need to be earnestly pondered. A settlement which, having destroyed the National Socialist rulers of Germany, leaves untouched the conditions which made the phenomenon of National Socialism possible, will run the risk of being as short-lived and as tragic as the settlement of 1919. No period of history will better repay study by the peacemakers of the future than the Twenty Years' Crisis which fills the interval between the two Great Wars. The next Peace Conference, if it is not to repeat the fiasco of the last, will have to concern itself with issues more fundamental than the drawing of frontiers. In this belief, I have ventured to dedicate this book to the makers of the coming peace.

The published sources from which I have derived help and inspiration are legion. I am specially indebted to two books which, though not specifically concerned with international relations, seem to me to have illuminated some of the fundamental problems of politics: Dr Karl Mannheim's *Ideology and Utopia* and Dr Reinhold Niebuhr's *Moral Man and Immoral Society*. Mr Peter Drucker's *The End of Economic Man*, which did not come into my hands until my manuscript was virtually complete, contains some brilliant guesses and a most stimulating and suggestive diagnosis of the present crisis in world history. Many excellent historical and descriptive works about various aspects of international relations have appeared in the last twenty years, and my indebtedness to some of these is recorded in endnotes, which must take the place of a bibliography. But not one of these works known to me has attempted to analyse the profounder causes of the contemporary international crisis.

My obligations to individuals are still more extensive. In particular, I desire to record my deep gratitude to three friends who found time to read the whole of my manuscript, whose comments were equally stimulating whether they agreed or disagreed with my views, and whose suggestions are responsible for a great part of such value as this book possesses: Charles Manning, Professor of International Relations in the London School of Economics and Political Science; Dennis Routh, Fellow of All Souls College, Oxford, and recently Lecturer in International Politics in the University College of Wales, Aberystwyth; and a third, whose official position deprives me of the pleasure of naming him here. During the past three years I have been a member of a Study Group of the Royal Institute of International Affairs engaged on an enquiry into the problem of nationalism, the results of which are about to be published.[1] The lines of investigation pursued by this Group have sometimes touched or crossed those which I have been following in these pages; and my colleagues in this Group and other contributors to its work have, in the course of our long discussions, unwittingly made numerous valuable contributions to the present book. To these, and to the many others who, in one way or another, consciously or unconsciously, have given me assistance and encouragement in the preparation of this volume, I tender my sincere thanks.

E. H. CARR

30 *September* 1939

---

[1] *Nationalism*: A study by a Group of Members of the Royal Institute of International Affairs (Oxford University Press).

Part One

The Science of International Politics

CHAPTER ONE

The Beginnings of a Science

The science of international politics is in its infancy. Down to 1914, the conduct of international relations was the concern of persons professionally engaged in it. In democratic countries, foreign policy was traditionally regarded as outside the scope of party politics; and the representative organs did not feel themselves competent to exercise any close control over the mysterious operations of foreign offices. In Great Britain, public opinion was readily aroused if war occurred in any region traditionally regarded as a sphere of British interest, or if the British navy momentarily ceased to possess that margin of superiority over potential rivals which was then deemed essential. In continental Europe, conscription and the chronic fear of foreign invasion had created a more general and continuous popular awareness of international problems. But this awareness found expression mainly in the labour movement, which from time to time passed somewhat academic resolutions against war. The constitution of the United States of America contained the unique provision that treaties were concluded by the President 'by and with the advice and consent of the Senate'. But the foreign relations of the United States seemed too parochial to lend any wider significance to this exception. The more picturesque aspects of diplomacy had a certain news value. But nowhere, whether in universities or in wider intellectual circles, was there organized study of current international affairs. War was still regarded mainly as the business of soldiers; and the corollary of this was that international politics were the business of diplomats. There was no general desire to take the conduct of international affairs out of the hands of the professionals or even to pay serious and systematic attention to what they were doing.

The war of 1914–18 made an end of the view that war is a matter which affects only professional soldiers and, in so doing, dissipated the corresponding impression that international politics could safely be left in the hands of professional diplomats. The campaign for the popularization of international politics began in the English-speaking countries in the form of an agitation against secret treaties, which were attacked, on insufficient evidence, as one of the causes of the war. The blame for the secret treaties should have been imputed, not to the wickedness of the governments, but to the indifference of the peoples. Everybody knew that such treaties were concluded. But before the war of 1914 few people felt any curiosity about them or thought them objectionable.[1] The agitation against them was, however, a fact of immense importance. It was the first symptom of the demand for the popularization of international politics and heralded the birth of a new science.

Purpose and analysis in political science

The science of international politics has, then, come into being in response to a popular demand. It has been created to serve a purpose and has, in this respect, followed the pattern of other sciences. At first sight, this pattern may appear illogical. Our first business, it will be said, is to collect, classify and analyse our facts and draw our inferences; and we shall then be ready to investigate the purpose to which our facts and our deductions can be put. The processes of the human mind do not, however, appear to develop in this logical order. The human mind works, so to speak, backwards. Purpose, which should logically follow analysis, is required to give it both its initial impulse and its direction. 'If society has a technical need,' wrote Engels, 'it serves as a greater spur to the progress of science than do ten universities.'[2] The first extant textbook of geometry 'lays down an aggregate of practical rules designed to solve concrete problems: "rule for measuring a round fruitery"; "rule for laying out a field"; "computation of the fodder consumed by geese and oxen"'.[3] Reason, says Kant, must approach nature 'not ... in the character of a pupil, who listens to all that his master chooses to tell him, but in that of a judge, who compels the witnesses to reply to those questions which he himself thinks fit to propose'.[4] 'We cannot study even stars or rocks or atoms', writes a modern sociologist, 'without being somehow determined, in our modes of systematization, in the prominence given to one or another part of our subject, in the form of the questions we ask and attempt to answer, by direct and human interests.'[5] It is the purpose of promoting health which creates medical science, and the purpose of building bridges which creates the science of engineering. Desire to cure the sicknesses of the body politic has given its impulse and its inspiration to political science. Purpose, whether we are conscious of it or not, is a condition of thought; and thinking for thinking's sake is as abnormal and barren as the miser's accumulation of money for its own sake. 'The wish is father to the thought' is a perfectly exact description of the origin of normal human thinking.

If this is true of the physical sciences, it is true of political science in a far more intimate sense. In the physical sciences, the distinction between the investigation of facts and the purpose to which the facts are to be put is not only theoretically valid, but is constantly observed in practice. The laboratory worker engaged in investigating the causes of cancer may have been originally inspired by the purpose of eradicating the disease. But this purpose is in the strictest sense irrelevant to the investigation and separable from it. His conclusion can be nothing more than a true report on facts. It cannot help to make the facts other than they are; for the facts exist independently of what anyone thinks about them. In the political sciences, which are concerned with human behaviour, there are no such facts. The investigator is inspired by the desire to cure some ill of the body politic. Among the causes of the trouble, he diagnoses the fact that human beings normally react to certain conditions in a certain way. But this is not a fact comparable with the fact that human bodies react in a certain way to certain drugs. It is a fact which may be changed by the desire to change it; and this desire, already present in the mind of the investigator, may be extended, as the result of his investigation, to a sufficient number of other human beings to make it effective. The purpose is not, as in the physical sciences, irrelevant to the investigation and separable from it: it is itself one of the facts. In theory, the distinction may no doubt still be drawn between the role of the investigator who establishes the facts and the role of the practitioner who considers the right course of action. In practice, one role shades imperceptibly into the other. Purpose and analysis become part and parcel of a single process.

A few examples will illustrate this point. Marx, when he wrote *Capital*, was inspired by the purpose of destroying the capitalist system just as the investigator of the causes of cancer is inspired by the purpose of eradicating cancer. But the facts about capitalism are not, like the facts about cancer, independent of the attitude of people towards it. Marx's analysis was intended to alter, and did in fact alter, that attitude. In the process of analysing the facts, Marx altered them. To attempt to distinguish between Marx the scientist and Marx the propagandist is idle hair-splitting. The financial experts, who in the summer of 1932 advised the British Government that it was possible to convert 5 per cent War Loan at the rate of 3½ per cent, no doubt based their advice on an analysis of certain facts; but the fact that they gave this advice was one of the facts which, being known to the financial world, made the operation successful. Analysis and purpose were inextricably blended. Nor is it only the thinking of professional or qualified students of politics which constitutes a political fact. Everyone who reads the political columns of a newspaper or attends a political meeting or discusses politics with his neighbour is to that extent a student of politics; and the judgement which he forms becomes (especially, but not exclusively, in democratic countries) a factor in the course of political events. Thus a reviewer might conceivably criticize this book on the ground, not that it was false, but that it was inopportune; and this criticism, whether justified or not, would be intelligible, whereas the same criticism of a book about the causes of cancer would be meaningless. Every political judgement helps to modify the facts on which it is passed. Political thought is itself a form of political action. Political science is the science not only of what is, but of what ought to be.

The role of utopianism

If therefore purpose precedes and conditions thought, it is not surprising to find that, when the human mind begins to exercise itself in some fresh field, an initial stage occurs in which the element of wish or purpose is overwhelmingly strong, and the inclination to analyse facts and means weak or non-existent. Hobhouse notes as a characteristic of 'the most primitive peoples' that 'the evidence of the truth of an idea is not yet separate from the quality which renders it pleasant'.[6] The same would appear to be conspicuously true of the primitive, or 'utopian', stage of the political sciences. During this stage, the investigators will pay little attention to existing 'facts' or to the analysis of cause and effect, but will devote themselves wholeheartedly to the elaboration of visionary projects for the attainment of the ends which they have in view – projects whose simplicity and perfection give them an easy and universal appeal. It is only when these projects break down, and wish or purpose is shown to be incapable by itself of achieving the desired end, that the investigators will reluctantly call in the aid of analysis, and the study, emerging from its infantile and utopian period, will establish its claim to be regarded as a science. 'Sociology', remarks Professor Ginsberg, 'may be said to have arisen by way of reaction against sweeping generalizations unsupported by detailed inductive enquiry.'[7]

It may not be fanciful to find an illustration of this rule even in the domain of physical science. During the Middle Ages, gold was a recognized medium of exchange. But economic relations were not sufficiently developed to require more than a limited amount of such a medium. When the new economic conditions of the fourteenth and fifteenth centuries introduced a widespread system of money transactions, and the supply of gold was found to be inadequate for the purpose, the wise men of the day began to experiment in the possibility of transmuting commoner metals into gold. The thought of the alchemist was purely purposive. He did not stop to enquire whether the properties of lead were such as to make it transmutable into gold. He assumed that the end was absolute (i.e. that gold must be produced), and that means and material must somehow be adapted to it. It was only when this visionary project ended in failure that the investigators were prompted to apply their thought to an examination of 'facts', i.e. the nature of matter; and though the initial utopian purpose of making gold out of lead is probably as far as ever from fulfilment, modern physical science has been evolved out of this primitive aspiration.

Other illustrations may be taken from fields more closely akin to our present subject.

It was in the fifth and fourth centuries BC that the first serious recorded attempts were made to create a science of politics. These attempts were made independently in China and in Greece. But neither Confucius nor Plato, though they were of course profoundly influenced by the political institutions under which they lived, really tried to analyse the nature of those institutions or to seek the underlying causes of the evils which they deplored. Like the alchemists, they were content to advocate highly imaginative solutions whose relation to existing facts was one of flat negation.[8] The new political order which they propounded was as different from anything they saw around them as gold from lead. It was the product not of analysis, but of aspiration.

In the eighteenth century, trade in Western Europe had become so important as to render irksome the innumerable restrictions placed on it by governmental authority and justified by mercantilist theory. The protest against these restrictions took the form of a wishful vision of universal free trade; and out of this vision the physiocrats in France, and Adam Smith in Great Britain, created a science of political economy. The new science was based primarily on a negation of existing reality and on certain artificial and unverified generalizations about the behaviour of a hypothetical economic man. In practice, it achieved some highly useful and important results. But economic theory long retained its utopian character; and even to-day some 'classical economists' insist on regarding universal free trade – an imaginary condition which has never existed – as the normal postulate of economic science, and all reality as a deviation from this utopian prototype.[9]

In the opening years of the nineteenth century, the industrial revolution created a new social problem to engage human thought in Western Europe. The pioneers who first set out to tackle this problem were the men on whom posterity has bestowed the name of 'utopian socialists': Saint-Simon and Fourier in France, Robert Owen in England. These men did not attempt to analyse the nature of class-interests or class-consciousness or of the class-conflict to which they gave rise. They simply made unverified assumptions about human behaviour and, on the strength of these, drew up visionary schemes of ideal communities in which men of all classes would live together in amity, sharing the fruits of their labours in proportion to their needs. For all of them, as Engels remarked, 'socialism is the expression of absolute truth, reason and justice, and needs only be discovered in order to conquer all the world in virtue of its own power'.[10] The utopian socialists did valuable work in making men conscious of the problem and of the need of tackling it. But the solution propounded by them had no logical connexion with the conditions which created the problem. Once more, it was the product not of analysis, but of aspiration.

Schemes elaborated in this spirit would not, of course, work. Just as nobody has ever been able to make gold in a laboratory, so nobody has ever been able to live in Plato's republic or in a world of universal free trade or in Fourier's phalansteries. But it is, nevertheless, perfectly right to venerate Confucius and Plato as the founders of political science, Adam Smith as the founder of political economy, and Fourier and Owen as the founders of socialism. The initial stage of aspiration towards an end is an essential foundation of human thinking. The wish is father to the thought. Teleology precedes analysis.

The teleological aspect of the science of international politics has been conspicuous from the outset. It took its rise from a great and disastrous war; and the overwhelming purpose which dominated and inspired the pioneers of the new science was to obviate a recurrence of this disease of the international body politic. The passionate desire to prevent war determined the whole initial course and direction of the study. Like other infant sciences, the science of international politics has been markedly and frankly utopian. It has been in the initial stage in which wishing prevails over thinking, generalization over observation, and in which little attempt is made at a critical analysis of existing facts or available means. In this stage, attention is concentrated almost exclusively on the end to be achieved. The end has seemed so important that analytical criticism of the means proposed has too often been branded as destructive and unhelpful. When President Wilson, on his way to the Peace Conference, was asked by some of his advisers whether he thought his plan of a League of Nations would work, he replied briefly: 'If it won't work, it must be made to work.'[11] The advocate of a scheme for an international police force or for 'collective security', or of some other project for an international order, generally replied to the critic not by an argument designed to show how and why he thought his plan will work, but either by a statement that it must be made to work because the consequences of its failure to work would be so disastrous, or by a demand for some alternative nostrum.[12] This must be the spirit in which the alchemist or the utopian socialist would have answered the sceptic who questioned whether lead could be turned into gold or men made to live in model communities. Thought has been at a discount. Much that was said and written about international politics between 1919 and 1939 merited the stricture applied in another context by the economist Marshall, who compares 'the nervous irresponsibility which conceives hasty utopian schemes' to the 'bold facility of the weak player who will speedily solve the most difficult chess problem by taking on himself to move the black men as well as the white'.[13] In extenuation of this intellectual failure, it may be said that, during the earlier of these years, the black pieces in international politics were in the hands of such weak players that the real difficulties of the game were scarcely manifest even to the keenest intelligence. The course of events after 1931 clearly revealed the inadequacy of pure aspiration as the basis for a science of international politics, and made it possible for the first time to embark on serious critical and analytical thought about international problems.

The impact of realism

No science deserves the name until it has acquired sufficient humility not to consider itself omnipotent, and to distinguish the analysis of what is from aspiration about what should be. Because in the political sciences this distinction can never be absolute, some people prefer to withhold from them the right to the title of science. In both physical and political sciences, the point is soon reached where the initial stage of wishing must be succeeded by a stage of hard and ruthless analysis. The difference is that political sciences can never wholly emancipate themselves from utopianism, and that the political scientist is apt to linger for a longer initial period than the physical scientist in the utopian stage of development. This is perfectly natural. For while the transmutation of lead into gold would be no nearer if everyone in the world passionately desired it, it is undeniable that if everyone really desired a 'world-state' or 'collective security' (and meant the same thing by those terms), it would be easily attained; and the student of international politics may be forgiven if he begins by supposing that his task is to make everyone desire it. It takes him some time to understand that no progress is likely to be made along this path, and that no political utopia will achieve even the most limited success unless it grows out of political reality. Having made the discovery, he will embark on that hard ruthless analysis of reality which is the hallmark of science; and one of the facts whose causes he will have to analyse is the fact that few people do desire a 'world-state' or 'collective security', and that those who think they desire it mean different and incompatible things by it. He will have reached a stage when purpose by itself is seen to be barren, and when analysis of reality has forced itself upon him as an essential ingredient of his study.

The impact of thinking upon wishing which, in the development of a science, follows the breakdown of its first visionary projects, and marks the end of its specifically utopian period, is commonly called realism. Representing a reaction against the wish-dreams of the initial stage, realism is liable to assume a critical and somewhat cynical aspect. In the field of thought, it places its emphasis on the acceptance of facts and on the analysis of their causes and consequences. It tends to depreciate the role of purpose and to maintain, explicitly or implicitly, that the function of thinking is to study a sequence of events which it is powerless to influence or to alter. In the field of action, realism tends to emphasize the irresistible strength of existing forces and the inevitable character of existing tendencies, and to insist that the highest wisdom lies in accepting, and adapting oneself to, these forces and these tendencies. Such an attitude, though advocated in the name of 'objective' thought, may no doubt be carried to a point where it results in the sterilization of thought and the negation of action. But there is a stage where realism is the necessary corrective to the exuberance of utopianism, just as in other periods utopianism must be invoked to counteract the barrenness of realism. Immature thought is predominantly purposive and utopian. Thought which rejects purpose altogether is the thought of old age. Mature thought combines purpose with observation and analysis. Utopia and reality are thus the two facets of political science. Sound political thought and sound political life will be found only where both have their place.

Notes

1. A recent historian of the Franco-Russian alliance, having recorded the protests of a few French radicals against the secrecy which enveloped this transaction, continues: 'Parliament and opinion tolerated this complete silence, and were content to remain in absolute ignorance of the provisions and scope of the agreement' (Michon, *L'Alliance Franco-Russe*, p. 75). In 1898, in the Chamber of Deputies, Hanotaux was applauded for describing the disclosure of its terms as 'absolutely impossible' (*ibid.*, p. 82).
2. Quoted in Sidney Hook, *Towards the Understanding of Karl Marx*, p. 279.
3. J. Rueff, *From the Physical to the Social Sciences* (Engl. transl.), p. 27.
4. Kant, *Critique of Pure Reason* (Everyman ed.), p. 11.
5. MacIver, *Community*, p. 56.
6. L. T. Hobhouse, *Development and Purpose*, p. 100.
7. M. Ginsberg, *Sociology*, p. 25.
8. 'Plato and Plotinus, More and Campanella constructed their fanciful societies with those materials which were omitted from the fabric of the actual communities by the defects of which they were inspired. The Republic, the Utopia, and the City of the Sun were protests against a state of things which the experience of their authors taught them to condemn' (Acton, *History of Freedom*, p. 270).

9. 'L'économie politique libérale a été un des meilleurs exemples d'utopies qu'on puisse citer. On avait imaginé une société où tout serait ramené à des types commerciaux, sous la loi de la plus complète concurrence; on reconnaît aujourd'hui que cette société idéale serait aussi difficile à realiser que celle de Platon' (Sorel, *Réflexions sur la violence*, p. 47). Compare Professor Robbins' well-known defence of *laissez-faire* economics: 'The idea of a co-ordination of human activity by means of a system of impersonal rules, within which what spontaneous relations arise are conducive to mutual benefit, is a conception at least as subtle, at least as ambitious, as the conception of prescribing each action or each type of action by a central planning authority; and it is perhaps not less in harmony with the requirements of a spiritually sound society' (*Economic Planning and International Order*, p. 229). It would be equally true, and perhaps equally useful, to say that the constitution of Plato's Republic is at least as subtle, ambitious and satisfying to spiritual requirements as that of any state which has ever existed.

10. Engels, *Socialism, Utopian and Scientific* (Engl. transl.), p. 26.

11. R. S. Baker, *Woodrow Wilson and World Settlement*, i. p. 93.

12. 'There is the old well-known story about the man who, during the Lisbon earthquake of 1775, went about hawking anti-earthquake pills; but one incident is forgotten – when someone pointed out that the pills could not possibly be of use, the hawker replied: 'But what would you put in their place?'' (L. B. Namier, *In the Margin of History*, p. 20).

13. *Economic Journal* (1907), xvii. p. 9.

CHAPTER TWO

Utopia and Reality

The antithesis of utopia and reality – a balance always swinging towards and away from equilibrium and never completely attaining it – is a fundamental antithesis revealing itself in many forms of thought. The two methods of approach – the inclination to ignore what was and what is in contemplation of what should be, and the inclination to deduce what should be from what was and what is – determine opposite attitudes towards every political problem. 'It is the eternal dispute', as Albert Sorel puts it, 'between those who imagine the world to suit their policy, and those who arrange their policy to suit the realities of the world.'[1] It may be suggestive to elaborate this antithesis before proceeding to an examination of the current crisis of international politics.

Free will and determination

The antithesis of utopia and reality can in some aspects be identified with the antithesis of Free Will and Determinism. The utopian is necessarily voluntarist: he believes in the possibility of more or less radically rejecting reality, and substituting his utopia for it by an act of will. The realist analyses a predetermined course of development which he is powerless to change. For the realist, philosophy, in the famous words of Hegel's preface to his *Philosophy of Right*, always 'comes too late' to change the world. By means of philosophy, the old order 'cannot be rejuvenated, but only known'. The utopian, fixing his eyes on the future, thinks in terms of creative spontaneity: the realist, rooted in the past, in terms of causality. All healthy human action, and therefore all healthy thought, must establish a balance between utopia and reality, between free will and determinism. The complete realist, unconditionally accepting the causal sequence of events, deprives himself of the possibility of changing reality. The complete utopian, by rejecting the causal sequence, deprives himself of the possibility of understanding either the reality which he is seeking to change or the processes by which it can be changed. The characteristic vice of the utopian is naïvety; of the realist, sterility.[2]

Theory and practice

The antithesis of utopia and reality also coincides with the antithesis of theory and practice. The utopian makes political theory a norm to which political practice ought to conform. The realist regards political theory as a sort of codification of political practice. The relationship of theory and practice has come to be recognized in recent years as one of the central problems of political thought. Both the utopian and the realist distort this relationship. The utopian, purporting to recognize the interdependence of purpose and fact, treats purpose as if it were the only relevant fact, and constantly couches optative propositions in the indicative mood. The American Declaration of Independence maintains that 'all men are created equal', Mr Litvinov that 'peace is indivisible',[3] and Sir Norman Angell that 'the biological division of mankind into independent warring states' is a 'scientific ineptitude'.[4] Yet it is a matter of common observation that all men are not born equal even in the United States, and that the Soviet Union can remain at peace while its neighbours are at war; and we should probably think little of a zoologist who described a man-eating tiger as a 'scientific ineptitude'. These propositions are items in a political pro-gramme disguised as statements of fact;[5] and the utopian inhabits a dream-world of such 'facts', remote from the world of reality where quite contrary facts may be observed. The realist has no difficulty in perceiving that these utopian propositions are not facts but aspirations, and belong to the optative not to the indicative mood; and he goes on to show that, considered as aspirations, they are not *a priori* propositions, but are rooted in the world of reality in a way which the utopian altogether fails to understand. Thus for the realist, the equality of man is the ideology of the underprivileged seeking to raise themselves to the level of the privileged; the indivisibility of peace the ideology of states which, being particularly exposed to attack, are eager to establish the principle that an attack on them is a matter of concern to other states more fortunately situated;[6] the ineptitude of sovereign states the ideology of predominant Powers which find the sovereignty of other states a barrier to the enjoyment of their own predominant position. This exposure of the hidden foundations of utopian theory is a necessary preliminary to any serious political science. But the realist, in denying any *a priori* quality to political theories, and in proving them to be rooted in practice, falls easily into a determinism which argues that theory, being nothing more than a rationalization of conditioned and predetermined purpose, is a pure excrescence and impotent to alter the course of events. While therefore the utopian treats purpose as the sole ultimate fact, the realist runs the risk of treating purpose merely as the mechanical product of other facts. If we recognize that this mechanization of human will and human aspiration is untenable and intolerable, then we must recognize that theory, as it develops out of practice and develops into practice, plays its own transforming role in the process. The political process does not consist, as the realist believes, purely in a succession of phenomena governed by mechanical laws of causation; nor does it consist, as the utopian believes, purely in the application to practice of certain theoretical truths evolved out of their inner consciousness by wise and far-seeing people. Political science must be based on a recognition of the interdependence of theory and practice, which can be attained only through a combination of utopia and reality.

The intellectual and the bureaucrat

A concrete expression of the antithesis of theory and practice in politics is the opposition between the 'intellectual' and the 'bureaucrat',[7] the former trained to think mainly on *a priori* lines, the latter empirically. It is in the nature of things that the intellectual should find himself in the camp which seeks to make practice conform to theory; for intellectuals are particularly reluctant to recognize their thought as conditioned by forces external to themselves, and like to think of themselves as leaders whose theories provide the motive force for so-called men of action. Moreover, the whole intellectual outlook of the last two hundred years has been strongly coloured by the mathematical and natural sciences. To establish a general principle, and to test the particular in the light of that principle, has been assumed by most intellectuals to be the necessary foundation and starting-point of any science. In this respect, utopianism with its insistence on general principles may be said to represent the characteristic intellectual approach to politics. Woodrow Wilson, the most perfect modern example of the intellectual in politics, 'excelled in the exposition of fundamentals. His political method ... was to base his appeal upon broad and simple principles, avoiding commitment upon specific measures.'[8] Some suppo-sedly general principle, such as 'national self-determination', 'free trade' or 'collective security' (all of which will be easily recognized by the realist as concrete expressions of particular conditions and interests), is taken as an absolute standard, and policies are adjudged good or bad by the extent to which they conform to, or diverge from, it. In modern times, intellectuals have been the leaders of every utopian movement; and the services which utopianism has rendered to political progress must be credited in large part to them. But the characteristic weakness of utopianism is also the characteristic weakness of the political intellectuals – failure to understand existing reality and the way in which their own standards are rooted in it. 'They could give to their political aspirations', wrote Meinecke of the role of intellectuals in German politics, 'a spirit of purity and independence, of philosophical idealism and of elevation above the concrete play of interests ... but through their defective feeling for the realistic interests of actual state life they quickly descended from the sublime to the extravagant and eccentric.'[9]

It has often been argued that the intellectuals are less directly conditioned in their thinking than those groups whose coherence depends on a common economic interest, and that they therefore occupy a vantage point *au-dessus de la mêlée*. As early as 1905, Lenin attacked 'the old-fashioned view of the intelligentsia as capable ... of standing outside class'.[10] More recently, this view has been resuscitated by Dr Mannheim, who argues that the intelligentsia, being '*relatively* classless' and 'socially unattached', 'subsumes in itself all those interests with which social life is permeated', and can therefore attain a higher measure of impartiality and objectivity.[11] In a certain limited sense, this is true. But any advantage derived from it would seem to be nullified by a corresponding disability, i.e. detachment from the masses whose attitude is the determining factor in political life. Even where the illusion of their leadership was strongest, modern intellectuals have often found themselves in the position of officers whose troops were ready enough to follow them in quiet times, but could be relied on to desert in any serious engagement. In Germany and many smaller European countries, the democratic constitutions of 1919 were the work of devoted intellectuals, and achieved a high degree of theoretical perfection. But when a crisis occurred, they broke down almost everywhere through failure to win the durable allegiance of the mass of the population. In the United States, the intellectuals played a preponderant part in creating the League of Nations, and most of them remained avowed supporters of it. Yet the mass of the American people, having appeared to follow their lead, rejected it when the critical moment arrived. In Great Britain, the intellectuals secured, by a devoted and energetic propaganda, overwhelming paper support for the League of Nations. But when the Covenant appeared to require action which might have entailed practical consequences for the mass of the people, successive governments preferred inaction; and the protests of the intellectuals caused no perceptible reaction in the country.

The bureaucratic approach to politics is, on the other hand, fundamentally empirical. The bureaucrat purports to handle each particular problem 'on its merits', to eschew the formulation of principles and to be guided on the right course by some intuitive process born of long experience and not of conscious reasoning. 'There are no general cases,' said a French official, acting as French Delegate at an Assembly of the League of Nations; 'there are only specific cases.'[12] In his dislike of theory, the bureaucrat resembles the man of action. 'On s'engage, puis on voit' is a motto attributed to more than one famous general. The excellence of the British civil service is partly due to the ease with which the bureaucratic mentality accommodates itself to the empirical tradition of British politics. The perfect civil servant conforms closely to the popular picture of the English politician as a man who recoils from written constitutions and solemn covenants, and lets himself be guided by precedent, by instinct, by feel for the right thing. This empiricism is itself, no doubt, conditioned by a specific point of view, and reflects the conservative habit of English political life. The bureaucrat, perhaps more explicitly than any other class of the community, is bound up with the existing order, the maintenance of tradition, and the acceptance of precedent as the 'safe' criterion of action. Hence bureaucracy easily degenerates into the rigid and empty formalism of the mandarin, and claims an esoteric understanding of appropriate procedures which is not accessible even to the most intelligent outsider. 'Expérience vaut mieux que science' is the typical bureaucratic motto. 'Attainments in learning and science', wrote Bryce, voicing a widely felt prejudice, 'do little to make men wise in politics.'[13] When a bureaucrat wishes to damn a proposal, he calls it 'academic'. Practice, not theory, bureaucratic training, not intellectual brilliance, is the school of political wisdom. The bureaucrat tends to make politics an end in themselves. It is worth remarking that both Machiavelli and Bacon were bureaucrats.

This fundamental antithesis between intellectual and bureaucratic modes of thought, always and everywhere latent, has appeared in the last half century in a quarter where it would hardly have been looked for: in the labour movement. Writing in the 1870s, Engels congratulated the German workers on the fact that they 'belong to the most theoretical nation in the world, and have retained that theoretical sense which has been almost completely lost by the so-called "educated" classes in Germany'. He contrasted this happy state with 'the indifference to all theory which is one of the chief reasons of the slow progress of the English workers' movement'.[14] Forty years later, another German writer confirmed this observation.[15] The theoretical analysis of Marxist doctrine became one of the principal preoccupations of leading German Social Democrats; and many observers believe that this one-sided intellectual development was an important factor in the ultimate collapse of the party. The British labour movement, until the last few years, entirely eschewed theory. At present, imperfect harmony between the intellectual and trade union wings is a notorious source of embarrassment to the Labour Party. The trade unionist tends to regard the intellectual as a utopian theorist lacking experience in the practical problems of the movement. The intellectual condemns the trade union leader as a bureaucrat. The recurrent conflicts between factions within the Bolshevik party in Soviet Russia were in part, at any rate, explicable as conflicts between the 'party intelligentsia', represented by Bukharin, Kamenev, Radek and Trotsky, and the 'party machine' represented by Lenin, Sverdlov (till his death in 1919) and Stalin.[16]

The opposition between intellectual and bureaucrat was particularly prominent in Great Britain during the twenty years between the wars in the field of foreign affairs. During the first world war, the Union of Democratic Control, an organization of utopian intellectuals, strove to popularize the view that the war was largely due to the control of foreign affairs in all countries by professional diplomats. Woodrow Wilson believed that peace would be secured if international issues were settled 'not by diplomats or politicians each eager to serve his own interests, but by dispassionate scientists – geographers, ethnologists, economists – who had made studies of the problems involved'.[17] Bureaucrats, and especially diplomats, were long regarded with suspicion in League of Nations circles; and it was considered that the League would contribute greatly to the solution of international problems by taking them out of the reactionary hands of foreign offices. Wilson, in introducing the draft Covenant to the plenary session of the Peace Conference, spoke of 'the feeling that, if the deliberating body of the League of Nations was merely to be a body of officials representing the various governments, the peoples of the world would not be sure that some of the mistakes which preoccupied officials had admittedly made might not be repeated'.[18] Later, in the House of Commons, Lord Cecil was more scathing:

I am afraid I came to the conclusion at the Peace Conference, from my own experience, that the Prussians were not exclusively confined to Germany. There is also the whole tendency and tradition of the official classes ... You cannot avoid the conclusion that there is a tendency among them to think that whatever is is right.[19]

At the Second Assembly, Lord Cecil invoked the support of 'public opinion', which the League was supposed to represent, against the 'official classes';[20] and such appeals were frequently heard during the next ten years. The bureaucrat for his part equally mistrusted the missionary zeal of enthusiastic intellectuals for collective security, world order and general disarmament – schemes which seemed to him the product of pure theory divorced from practical experience. The disarmament issue well illustrated this divergence of view. For the intellectual, the general principle was simple and straightforward; the alleged difficulties of applying it were due to obstruction by the 'experts'.[21] For the expert, the general principle was meaningless and utopian; whether armaments could be reduced, and if so which, was a 'practical' question to be decided in each case 'on its merits'.

Left and right

The antithesis of utopia and reality, and of theory and practice, further reproduces itself in the antithesis of radical and conservative, of Left and Right, though it would be rash to assume that parties carrying these labels always represent these underlying tendencies. The radical is necessarily utopian, and the conservative realist. The intellectual, the man of theory, will gravitate towards the Left just as naturally as the bureaucrat, the man of practice, will gravitate towards the Right. Hence the Right is weak in theory, and suffers through its inaccessibility to ideas. The characteristic weakness of the Left is failure to translate its theory into practice – a failure for which it is apt to blame the bureaucrats, but which is inherent in its utopian character. 'The Left has reason (*Vernunft*), the Right has wisdom (*Verstand*),' wrote the Nazi philosopher, Moeller van den Bruck.[22] From the days of Burke onwards, English conservatives have always strongly denied the possibility of deducing political practice by a logical process from political theory. 'To follow the syllogism alone is a short cut to the bottomless pit,' says Lord Baldwin[23] – a phrase which may suggest that he practices as well as preaches abstention from rigorously logical modes of thought. Mr Churchill refuses to believe that 'extravagant logic in doctrine' appeals to the British elector.[24] A particularly clear definition of different attitudes towards foreign policy comes from a speech made in the House of Commons by Neville Chamberlain in answer to a Labour critic:

> What does the hon. Member mean by foreign policy? You can lay down sound and general propositions. You can say that your foreign policy is to maintain peace; you can say that it is to protect British interests, you can say that it is to use your influence, such as it is, on behalf of the right against the wrong, as far as you can tell the right from the wrong. You can lay down all these general principles, but that is not a policy. Surely, if you are to have a policy you must take the particular situations and consider what action or inaction is suitable for those particular situations. That is what I myself mean by policy, and it is quite clear that as the situations and conditions in foreign affairs continually change from day to day, your policy cannot be stated once and for all, if it is to be applicable to every situation that arises.[25]

The intellectual superiority of the Left is seldom in doubt. The Left alone thinks out principles of political action and evolves ideals for statesmen to aim at. But it lacks practical experience which comes from close contact with reality. In Great Britain after 1919, it was a serious misfortune that the Left, having enjoyed office for negligible periods, had little experience of administrative realities and became more and more a party of pure theory, while the Right, having spent so little time in opposition, had few temptations to pit the perfection of theory against the imperfections of practice. In Soviet Russia, the group in power is more and more discarding theory in favour of practice as it loses the memory of its revolutionary origin. History everywhere shows that, when Left parties or politicians are brought into contact with reality through the assumption of political office, they tend to abandon their 'doctrinaire' utopianism and move towards the Right, often retaining their Left labels and thereby adding to the confusion of political terminology.

Ethics and politics

Most fundamental of all, the antithesis of utopia and reality is rooted in a different conception of the relationship of politics and ethics. The antithesis between the world of value and the world of nature, already implicit in the dichotomy of purpose and fact, is deeply embedded in the human consciousness and in political thought. The utopian sets up an ethical standard which purports to be independent of politics, and seeks to make politics conform to it. The realist cannot logically accept any standard value save that of fact. In his view, the absolute standard of the utopian is conditioned and dictated by the social order, and is therefore political. Morality can only be relative, not universal. Ethics must be interpreted in terms of politics; and the search for an ethical norm outside politics is doomed to frustration. The identification of the supreme reality with the supreme good, which Christianity achieves by a bold stroke of dogmatism, is achieved by the realist through the assumption that there is no good other than the acceptance and understanding of reality.

These implications of the opposition between utopia and reality will emerge clearly from a more detailed study of the modern crisis in international politics.

Notes

1. A. Sorel, *L'Europe et la Révolution Française*, p. 474.

2. The psychologist may be interested to trace here an analogy – it would be dangerous to treat it as more – with Jung's classification of psychological types as 'introverted' and 'extraverted' (Jung, *Psychological Types*) or William James's pairs of opposites: Rationalist-Empiricist, Intellectualist-Sensationalist, Idealist-Materialist, Optimistic-Pessimistic, Religious-Irreligious, Free-willist-Fatalistic, Monistic-Pluralistic, Dogmatical-Sceptical (W. James, *Pragmatism*).

3. *League of Nations: Sixteenth Assembly*, p. 72.

4. Angell, *The Great Illusion*, p. 138.

5. Similarly, Marx's theory of surplus value has, in the words of a sympathetic critic, 'rather the significance of a political and social slogan than of an economic truth' (M. Beer, *The Life and Teaching of Karl Marx*, p. 129).

6. Having discovered that other states were perhaps more open to attack than themselves, the Soviet authorities in May 1939 dismissed Mr Litvinov and ceased to talk about the indivisibility of peace.

7. The term 'bureaucrat' may be taken for this purpose to include those members of the fighting services who are concerned with the direction of policy. It is, perhaps, unnecessary to add that not every possessor of an intellect is an intellectual, or every occupant of a desk in a government department a bureaucrat. There are, nevertheless, modes of thought which are, broadly speaking, characteristic of the 'bureaucrat' and the 'intellectual' respectively.

8. R. S. Baker, *Woodrow Wilson: Life and Letters*, iii. p. 90.

9. Meinecke, *Staat und Persönlichkeit*, p. 136.

10. Lenin, *Works* (2nd Russian ed.), vii. p. 72.

11. Mannheim, *Ideology and Utopia*, pp. 137–40.

12. *League of Nations: Fifteenth Assembly*, Sixth Committee, p. 62.

13. Bryce, *Modern Democracies*, i. p. 89.

14. Quoted in Lenin, *Works* (2nd Russian ed.), iv. p. 381.

15. 'We possess the most theoretical labour movement in the world' (F. Naumann, *Central Europe*, Engl. transl., p. 121).

16. This interpretation, which appears in Mirsky's *Lenin* (pp. III, 117–18) published in 1931, received further confirmation from subsequent events. The rift went back to the earliest days of the party. In 1904 Lenin was attacking the party intellectuals for their neglect of discipline and organization, and the intellectuals were attacking Lenin's bureaucratic methods (Lenin, *Works* (2nd Russian ed.), vi. pp. 309–11).

17. R. S. Baker, *Woodrow Wilson and World Settlement*, i. p. 112.

18. *History of the Peace Conference*, ed. H. Temperley, iii. p. 62.

19. House of Commons, 21 July 1919: *Official Report*, col. 993.

20. *League of Nations: Second Assembly*, Third Committee, p. 281.

21. 'It is not to be tolerated', said the Belgian socialist De Brouckère, 'that the people should be robbed of their hopes of peace by experts who are losing themselves in the maze of technicalities which, with a little goodwill, might be disentangled in a few hours' (Peace and Disarmament Committee of the Women's International Organisations: Circular of 15 May 1932). About the same time, Lord Cecil was reported in the same sense: 'If the matter was to be

left to experts nothing would be done. They were, he was sure, most able, conscientious, highly instructed gentlemen, but just look at their training' (*Manchester Guardian*, 18 May 1932).

22. Moeller van den Bruck, *Das Dritte Reich* (3rd ed.), p. 257.
23. Baldwin, *On England*, p. 153.
24. Winston Churchill, *Step by Step*, p. 147.
25. House of Commons, 21 October 1937, reprinted in N. Chamberlain, *The Struggle for Peace*, p. 33.

Part Two

The International Crisis

CHAPTER THREE

The Utopian Background

The foundations of utopianism

The modern school of utopian political thought must be traced back to the break-up of the mediaeval system, which presupposed a universal ethic and a universal political system based on divine authority. The realists of the Renaissance made the first determined onslaught on the primacy of ethics and propounded a view of politics which made ethics an instrument of politics, the authority of the state being thus substituted for the authority of the church as the arbiter of morality. The answer of the utopian school to this challenge was not an easy one. An ethical standard was required which would be independent of any external authority, ecclesiastical or civil; and the solution was found in the doctrine of a secular 'law of nature' whose ultimate source was the individual human reason. Natural law, as first propounded by the Greeks, had been an intuition of the human heart about what is morally right. 'It is eternal', said Sophocles' Antigone, 'and no man knows whence it came.' The Stoics and the mediaeval schoolmen identified natural law with reason; and in the seventeenth and eighteenth centuries this identification was revived in a new and special form. In science, the laws of nature were deduced by a process of reasoning from observed facts about the nature of matter. By an easy analogy, the Newtonian principles were now applied to the ethical problems. The moral law of nature could be scientifically established; and rational deduction from the supposed facts of human nature took the place of revelation or intuition as the source of morality. Reason could determine what were the universally valid moral laws; and the assumption was made that, once these laws were determined, human beings would conform to them just as matter conformed to the physical laws of nature. Enlightenment was the royal road to the millennium.

By the eighteenth century, the main lines of modern utopian thought were firmly established. It was essentially individualist in that it made the human conscience the final court of appeal in moral questions; in France it became associated with a secular, in England with an evangelical tradition. It was essentially rationalist in that it identified the human conscience with the voice of reason.[1] But it had still to undergo important developments; and it was Jeremy Bentham who, when the industrial revolution had transferred the leadership of thought from France to England, gave to nineteenth-century utopianism its characteristic shape. Starting from the postulate that the fundamental characteristic of human nature is to seek pleasure and avoid pain, Bentham deduced from this postulate a rational ethic which defined the good in the famous formula 'the greatest happiness of the greatest number'. As has often been pointed out, 'the greatest happiness of the greatest number' performed the function, which natural law had performed for a previous generation, of an absolute ethical standard. Bentham firmly believed in this absolute standard, and rejected as 'anarchical' the view that there are 'as many standards of right and wrong as there are men'.[2] In effect, 'the greatest happiness of the greatest number' was the nineteenth-century definition of the content of natural law.

The importance of Bentham's contribution was twofold. In the first place, by identifying the good with happiness, he provided a plausible confirmation of the 'scientific' assumption of the eighteenth-century rationalists that man would infallibly conform to the moral law of nature once its content had been rationally determined. Secondly, while preserving the rationalist and individualist aspect of the doctrine, he succeeded in giving it a broader basis. The doctrine of reason in its eighteenth-century guise was pre-eminently intellectual and aristocratic. Its political corollary was an enlightened despotism of philosophers, who alone could be expected to have the necessary reasoning power to discover the good. But now that happiness was the criterion, the one thing needful was that the individual should understand where his happiness lay. Not only was the good ascertainable – as the eighteenth century had held – by a rational process, but this process – added the nineteenth century – was not a matter of abstruse philosophical speculation, but of simple common sense. Bentham was the first thinker to elaborate the doctrine of salvation by public opinion. The members of the community 'may, in their aggregate capacity, be considered as constituting a sort of judicatory or tribunal – call it ... *The Public-Opinion Tribunal*'.[3] It was James Mill, Bentham's pupil, who produced the most complete argument yet framed for the infallibility of public opinion:

Every man possessed of reason is accustomed to weigh evidence and to be guided and determined by its preponderance. When various conclusions are, with their evidence, presented with equal care and with equal skill, there is a moral certainty, though some few may be misguided, that the greatest number will judge right, and that the greatest force of evidence, whatever it is, will produce the greatest impression.[4]

This is not the only argument by which democracy as a political institution can be defended. But this argument was, in fact, explicitly or implicitly accepted by most nineteenth-century liberals. The belief that public opinion can be relied on to judge rightly on any question rationally presented to it, combined with the assumption that it will act in accordance with this right judgement, is an essential foundation of the liberal creed. In Great Britain, the later eighteenth and the nineteenth centuries were pre-eminently the age of popular preaching and of political oratory. By the voice of reason men could be persuaded both to save their own immortal souls and to move along the path of political enlightenment and progress. The optimism of the nineteenth century was based on the triple conviction that the pursuit of the good was a matter of right reasoning, that the spread of knowledge would soon make it possible for everyone to reason rightly on this important subject, and that anyone who reasoned rightly on it would necessarily act rightly.

The application of these principles to international affairs followed, in the main, the same pattern. The Abbé Saint-Pierre, who propounded one of the earliest schemes for a League of Nations, 'was so confident in the reasonableness of his projects that he always believed that, if they were fairly considered, the ruling powers could not fail to adopt them'.[5] Both Rousseau and Kant argued that, since wars were waged by princes in their own interest and not in that of their peoples, there would be no wars under a republican form of government. In this sense, they anticipated the view that public opinion, if allowed to make itself effective, would suffice to prevent war. In the nineteenth century, this view won widespread approval in Western Europe, and took on the specifically rationalist colour proper to the doctrine that the holding of the right moral beliefs and the performance of the right actions can be assured by process of reasoning. Never was there an age which so unreservedly proclaimed the supremacy of the intellect. 'It is intellectual evolution', averred Comte, 'which essentially determines the main course of social phenomena.'[6] Buckle, whose famous *History of Civilisation* was published between 1857 and 1861, boldly declared that dislike of war is 'a cultivated taste peculiar to an intellectual people'. He chose a cogent example, based on the assumption, natural to a British thinker, of the ingrained bellicosity of Great Britain's most recent enemy. 'Russia is a warlike country', he wrote, 'not because the inhabitants are immoral, but because they are unintellectual. The fault is in the head, not in the heart.'[7] The view that the spread of education would lead to international peace was shared by many of Buckle's contemporaries and successors. Its last serious exponent was Sir Norman Angell, who sought, by *The Great Illusion* and other books, to convince the world that war never brought profit to anyone. If he could establish this point by irrefutable argument, thought Sir Norman, then war could not occur. War was simply a 'failure of understanding'. Once the head was purged of the illusion that war was profitable, the heart could look after itself. 'The world of the Crusades and of heretic burning', ran the opening manifesto of a monthly journal called *War and Peace* which started publication in October 1913, '... was not a badly-meaning, but a badly-thinking world ... We emerged from it by correcting a defect in understanding; we shall emerge from the world of political warfare or armed peace in the same way.'[8] Reason could demonstrate the absurdity of the international anarchy; and with increasing knowledge, enough people would be rationally convinced of its absurdity to put an end to it.

Benthamism transplanted

Before the end of the nineteenth century, serious doubts had been thrown from more than one quarter on the assumptions of Benthamite rationalism. The belief in the sufficiency of reason to promote right conduct was challenged by psychologists. The identification of virtue with enlightened self-interest began to shock philosophers. The belief in the infallibility of public opinion had been attractive on the hypothesis of the earlier utilitarians that public opinion was the opinion of educated and enlightened men. It was less attractive, at any rate to those who thought themselves educated and enlightened, now that public opinion was the opinion of the masses; and as early as 1859, in his essay *On Liberty*, J. S. Mill had been preoccupied with the dangers of 'the tyranny of the majority'. After 1900, it would have been difficult to find, either in Great Britain or in any other European country, any serious political thinker who accepted the Benthamite assumptions without qualification. Yet, by one of the ironies of history, these half-discarded nineteenth-century assumptions reappeared, in the second and third decades of the twentieth century, in the special field of international politics, and there became the foundation-stones of a new utopian edifice. The explanation may be in part that, after 1914, men's minds naturally fumbled their way back, in search of a new utopia, to those apparently firm foundations of nineteenth-century peace and security. But a more decisive factor was the influence of the United States, still in the heyday of Victorian prosperity and of Victorian belief in the comfortable Benthamite creed. Just as Bentham, a century earlier, had taken the eighteenth-century doctrine of reason and refashioned it to the needs of the coming age, so now Woodrow Wilson, the impassioned admirer of

Bright and Gladstone, transplanted the nineteenth-century rationalist faith to the almost virgin soil of international politics and, bringing it back with him to Europe, gave it a new lease of life. Nearly all popular theories of international politics between the two world wars were reflexions, seen in an American mirror, of nineteenth-century liberal thought.

In a limited number of countries, nineteenth-century liberal democracy had been a brilliant success. It was a success because its presuppositions coincided with the stage of development reached by the countries concerned. Out of the mass of current speculation, the leading spirits of the age took precisely that body of theory which corresponded to their needs, consciously and unconsciously fitting their practice to it, and it to their practice. Utilitarianism and *laissez-faire* served, and in turn directed, the course of industrial and commercial expansion. But the view that nineteenth-century liberal democracy was based, not on a balance of forces peculiar to the economic development of the period and the countries concerned, but on certain *a priori* rational principles which had only to be applied in other contexts to produce similar results, was essentially utopian; and it was this view which, under Wilson's inspiration, dominated the world after the first world war. When the theories of liberal democracy were transplanted, by a purely intellectual process, to a period and to countries whose stage of development and whose practical needs were utterly different from those of Western Europe in the nineteenth century, sterility and disillusionment were the inevitable sequel. Rationalism can create a utopia, but cannot make it real. The liberal democracies scattered throughout the world by the peace settlement of 1919 were the product of abstract theory, stuck no roots in the soil, and quickly shrivelled away.

Rationalism and the League of Nations

The most important of all the institutions affected by this one-sided intellectualism of international politics was the League of Nations, which was an attempt 'to apply the principles of Lockeian liberalism to the building of a machinery of international order'.[9] 'The Covenant', observed General Smuts, '... simply carries into world affairs that outlook of a liberal democratic society which is one of the great achievements of our human advance.'[10] But this transplantation of democratic rationalism from the national to the international sphere was full of unforeseen difficulties. The empiricist treats the concrete case on its individual merits. The rationalist refers it to an abstract general principle. Any social order implies a large measure of standardization, and therefore of abstraction; there cannot be a different rule for every member of the community. Such standardization is comparatively easy in a community of several million anonymous individuals conforming more or less closely to recognized types. But it presents infinite complications when applied to sixty known states differing widely in size, in power, and in political, economic and cultural development. The League of Nations, being the first large-scale attempt to standardize international political problems on a rational basis, was particularly liable to these embarrassments.

The founders of the League, some of whom were men of political experience and political understanding, had indeed recognized the dangers of abstract perfection. 'Acceptance of the political facts of the present', remarked the official British Commentary on the Covenant issued in 1919, 'has been one of the principles on which the Commission has worked',[11] and this attempt to take account of political realities distinguished the Covenant not only from previous paper schemes of world organization, but also from such purely utopian projects as the International Police Force, the Kellogg–Briand Pact and the United States of Europe. The Covenant possessed the virtue of several theoretical imperfections. Purporting to treat all members as equal, it assured to the Great Powers a permanent majority on the Council of the League.[12] It did not purport to prohibit war altogether, but only to limit the occasions on which it might legitimately be resorted to. The obligation imposed on members of the League to apply sanctions to the Covenant-breaker was not free from vagueness; and this vagueness had been discreetly enhanced by a set of 'interpretative' resolutions passed by the Assembly of 1921. The starkness of the territorial guarantee provided by Article 10 of the Covenant was smoothed away in a resolution which secured an almost unanimous vote at the Assembly of 1923. It seemed for the moment as if the League might reach a working compromise between utopia and reality and become an effective instrument of international politics.

Unhappily, the most influential European politicians neglected the League during its critical formative years. Abstract rationalism gained the upper hand, and from about 1922 onwards the current at Geneva set strongly in the utopian direction.[13] It came to be believed, in the words of an acute critic, 'that there can exist, either at Geneva or in foreign offices, a sort of carefully classified card-index of events or, better still, "situations", and that, when the event happens or the situation presents itself, a member of the Council or Foreign Minister can easily recognize that event or situation and turn up the index to be directed to the files where the appropriate action is prescribed'.[14] There were determined efforts to perfect the machinery, to standardize the procedure, to close the 'gaps' in the Covenant by an absolute veto on all war, and to make the application of sanctions 'automatic'. The Draft Treaty of Mutual Assistance, the

Geneva Protocol, the General Act, the plan to incorporate the Kellogg–Briand Pact in the Covenant and 'the definition of the aggressor', were all milestones on the dangerous path of rationalization. The fact that the utopian dishes prepared during these years at Geneva proved unpalatable to most of the principal governments concerned was a symptom of the growing divorce between theory and practice.

Even the language current in League circles betrayed the growing eagerness to avoid the concrete in favour of the abstract generalizations. When it was desired to arrange that the Draft Treaty of Mutual Assistance could be brought into force in Europe without waiting for the rest of the world, a stipulation was inserted that it might come into force 'by continents' – a proviso with farcical implications for every continent except Europe. A conventional phraseology came into use, which served as the current coin of delegates at Geneva and of League enthusiasts elsewhere and which, through constant repetition, soon lost all contact with reality. 'I cannot recall any time,' said Mr Churchill in 1932, 'when the gap between the kind of words which statesmen used and what was actually happening in many countries was so great as it is now.'[15] The Franco-Soviet Pact, which was a defensive alliance against Germany, was so drafted as to make it appear an instrument of general application, and was described as a shining example of the principle of 'collective security'. A member of the House of Commons, when asked in the debate on sanctions in June 1936 whether he would run the risk of war with Italy, replied that he was prepared to face 'all the consequences naturally flowing from the enforcement of the Covenant against an aggressor nation'.[16] These linguistic contortions encouraged the frequent failure to distinguish between the world of abstract reason and the world of political reality. 'Metaphysicians, like savages,' remarks Mr Bertrand Russell, 'are apt to imagine a magical connexion between words and things'.[17] The metaphysicians of Geneva found it difficult to believe that an accumulation of ingenious texts prohibiting war was not a barrier against war itself. 'Our purpose', said M. Benes in introducing the Geneva Protocol to the 1924 Assembly, 'was to make war impossible, to kill it, to annihilate it. To do this we had to create a system.'[18] The Protocol was the 'system'. Such presumption could only provoke nemesis. Once it came to be believed in League circles that salvation could be found in a perfect card-index, and that the unruly flow of international politics could be canalized into a set of logically impregnable abstract formulae inspired by the doctrines of nineteenth-century liberal democracy, the end of the League as an effective political instrument was in sight.

The apotheosis of public opinion

Nor did any better fortune attend the attempt to transplant to the international sphere the liberal democratic faith in public opinion. And here there was a double fallacy. The nineteenth-century belief in public opinion comprised two articles: first (and in democracies this was, with some reservations, true), that public opinion is bound in the long run to prevail; and second (this was the Benthamite view), that public opinion is always right. Both these beliefs, not always clearly distinguished one from the other, were uncritically reproduced in the sphere of international politics.

The first attempts to invoke public opinion as a force in the international world had been made in the United States. In 1909, President Taft evolved a plan for the conclusion of treaties between the United States and other Great Powers for the compulsory arbitration of international disputes. But how, it was asked, would the award of the arbitral court be enforced? Taft disposed of the question with complete light-heartedness. He had never observed that in a democracy like the United States the enforcement of awards gave rise to any particular difficulty; and he professed himself 'very little concerned' about this aspect of the matter. 'After we have gotten the cases into court and decided, and the judgments embodied in a solemn declaration of a court thus established, few nations will care to face the condemnation of international public opinion and disobey the judgment.'[19] Public opinion, as in democratic countries, was bound to prevail; and public opinion, as the Benthamites said, could always be trusted to come down on the right side. The United States Senate rejected the President's proposal, so that the opportunity did not occur to put 'international public opinion' to the test. Four years later, Bryan, Wilson's first Secretary of State, came forward with a further set of treaties. In the Bryan treaties, arbitration was dropped in favour of conciliation. Their most novel and significant feature was the provision that the parties to them should not resort to war until twelve months had elapsed from the beginning of the dispute. In hot blood, the Bryan treaties seemed to admit, men might not listen to the voice of reason. But once delay had cooled their passions, reason, in the guise of international public opinion, would resume her compelling force. Several such treaties were in fact signed between the United States and other Powers – some of them, by a curious irony, in the first days of the first world war. 'The sum and substance' of these treaties, said Wilson in October 1914, was 'that whenever any trouble arises the light shall shine on it for a year before anything is done; and my prediction is that after the light has shone on it for a year, it will not be necessary to do anything; that after we know what happened, then we will know who was right and who was wrong.'[20]

The belief in the compelling power of reason, expressed through the voice of the people, was particularly congenial to Wilson. When he entered politics in 1910 as a candidate for the Governorship of New Jersey, his campaign was based on an appeal to 'the people' against the political bosses; and he displayed 'an almost mystical faith that the people would follow him if he could speak to enough of them'. The result of his campaign confirmed him in his belief in the potency of the voice of reason speaking through his lips. He would govern by the persuasiveness of reason acting on an all-powerful public opinion. 'If the bosses held back, he had only to appeal to the people. . . . The people wanted the high things, the right things, the true things.'[21]

America's entry into the war entailed no modification of Wilson's faith in the rightness of popular judgement. He took up the cue in one of the speeches in which he discussed the future conditions of peace:

> It is the peculiarity of this great war that, while statesmen have seemed to cast about for definitions of their purpose and have sometimes seemed to shift their ground and their point of view, the thought of the mass of men, whom statesmen are supposed to instruct and lead, has grown more and more unclouded, more and more certain of what it is they are fighting for. National purposes have fallen more and more into the background; and the common purpose of enlightened mankind has taken their place. The counsels of plain men have become on all hands more simple and straightforward and more unified than the counsels of sophisticated men of affairs, who still retain the impression that they are playing a game of power and are playing for high stakes. That is why I have said that this is a people's war, not a statesmen's. Statesmen must follow the clarified common thought or be broken.[22]

'Unless the Conference was prepared to follow the opinions of mankind,' he said on his way to Paris, 'and to express the will of the people rather than that of the leaders of the Conference, we should be involved in another break-up of the world.'[23]

Such conceptions did, in fact, play a conspicuous part in the work of the Conference. When the Italian Delegates proved recalcitrant in their claims to Fiume and the Adriatic coast, Wilson remained convinced that if he could appeal against the 'leaders' to the 'people', if only (as at the New Jersey election) he 'could speak to enough of them', the voice of reason must infallibly prevail. The communiqué to the Italian people, and the withdrawal of the Italian Delegation from Paris, were the result of this conviction. The problem of disarmament was approached in the same spirit. Once the enemy Powers had been disarmed by force, the voice of reason, speaking through public opinion, could be trusted to disarm the Allies. Both Wilson and Mr Lloyd George 'felt that, if the German army was limited, France would have to follow suit, and that she could hardly maintain an immense army under those conditions'.[24] And if anyone had paused to enquire on what compulsion France would have to disarm, the only answer could have been an appeal to the rational force of public opinion. Most important of all, the whole conception of the League of Nations was from the first closely bound up with the twin belief that public opinion was bound to prevail and that public opinion was the voice of reason. If 'open covenants openly arrived at' could be made the rule, the plain people could be relied on to see that the contents conformed to the requirements of that reason which was the highest morality. The new order must be based, not on 'covenants of selfishness and compromise' between governments, but on 'the thought of the plain people here and everywhere throughout the world, the people who enjoy no privilege and have very simple and unsophisticated standards of right and wrong'.[25] It must be 'sustained by the organized opinion of mankind'.[26]

The ticklish problem of material sanctions was approached reluctantly from the American, and almost as reluctantly from the British, side. Like Taft, Anglo-Saxon opinion felt itself 'very little concerned' over this aspect of the matter; for the recognition of the necessity of sanctions was in itself a derogation from the utopian doctrine of the efficacy of rational public opinion. It was unthinkable that a unanimous verdict of the League should be defied; and even if by some mischance the verdict were not unanimous, 'a majority report would probably be issued, and ... this', suggested Lord Cecil during the debates in Paris, 'would be likely to carry great weight with the public opinion of the world'.[27] The official British Commentary on the Covenant developed the same point of view:

> The League [it declared] must continue to depend on the free consent, in the last resort, of its component States; this assumption is evident in nearly every article of the Covenant, of which the ultimate and most effective sanction must be the public opinion of the civilized world. If the nations of the future are in the main selfish, grasping and warlike, no instrument or machinery will restrain them. It is only possible to establish an organization which may make peaceful co-operation easy and hence customary, and to trust in the influence of custom to mould public opinion.

The sanctions provisions were slurred over, half apologetically and with a consolatory postscript:

Not the least important part of the pressure will be supplied by the publicity stipulated for in the procedure of settlement. The obscure issues from which international quarrels arise will be dragged out into the light of day and the creation of an informed public opinion made possible.[28]

When the House of Commons debated the ratification of the Versailles Treaty, Lord Cecil was the principal expositor of the League Covenant:

For the most part [he told the House] there is no attempt to rely on anything like a superstate; no attempt to rely upon force to carry out a decision of the Council or the Assembly of the League. That is almost impracticable as things stand now. What we rely upon is public opinion ... and if we are wrong about it, then the whole thing is wrong.[29]

Addressing the Imperial Conference of 1923 on the subject of the League, Lord Cecil explained that 'its method is not ... the method of coercive government: it is a method of consent and its executive instrument is not force, but public opinion'.[30] And when the first League Assembly met, Lord Cecil, as British Delegate, propounded the same philosophy from the tribune:

It is quite true that by far the most powerful weapon at the command of the League of Nations is not the economic or the military weapon or any other weapon of material force. By far the strongest weapon we have is the weapon of public opinion.[31]

Even the more sceptical and sophisticated Balfour, explaining the absence of sanctions from the Washington agreements of 1921, declared that 'if any nation hereafter deliberately separates itself from the collective action we have taken in Washington in this year of grace, it will stand condemned before the world';[32] and it was one of the presuppositions of liberal democracy that such condemnation would be effective. But the argument that public opinion is the all-important weapon is two-edged; and in 1932, during the Manchurian crisis, the ingenious Sir John Simon used it to demonstrate that any other kind of action was superfluous. 'The truth is', he told the House of Commons, 'that when public opinion, world opinion, is sufficiently unanimous to pronounce a firm moral condemnation, sanctions are not needed.'[33] Given the Benthamite and Wilsonian premises, this answer was irrefutable. If public opinion had failed to curb Japan, then – as Lord Cecil had said in 1919 – 'the whole thing is wrong'.

The nemesis of utopianism in international politics came rather suddenly. In September 1930, the President of Columbia University, Dr Nicholas Murray Butler, ventured on the 'reasonably safe prediction that the next generation will see a constantly increasing respect for Cobden's principles and point of view and a steadily growing endeavour more largely to give them practical effect in public policy'.[34] On 10 September 1931, Lord Cecil told the Assembly of the League of Nations that 'there has scarcely ever been a period in the world's history when war seems less likely than it does at present'.[35] On 18 September 1931, Japan opened her campaign in Manchuria; and in the following month, the last important country which had continued to adhere to the principle of free trade took the first steps towards the introduction of a general tariff.

From this point onwards, a rapid succession of events forced upon all serious thinkers a reconsideration of premises which were becoming more and more flagrantly divorced from reality. The Manchurian crisis had demonstrated that the 'condemnation of international public opinion', invoked by Taft and by so many after him, was a broken reed. In the United States, this conclusion was drawn with extreme reluctance. In 1932, an American Secretary of State still cautiously maintained that 'the sanction of public opinion can be made one of the most potent sanctions of the world'.[36] In September 1938, President Roosevelt based his intervention in the Czecho-Slovak crisis on the belief of the United States Government in 'the moral force of public opinion',[37] and in April 1939, Mr Cordell Hull once again announced the conviction that 'a public opinion, the most potent of all forces for peace, is more strongly developing throughout the world'.[38] But in countries more directly menaced by international crisis, this consoling view no longer found many adherents; and the continued addition to it of American statesmen was regarded as an index of American unwillingness to resort to more potent weapons. Already in 1932, Mr Churchill taunted the League of Nations Union with 'long-suffering and inexhaustible gullibility' for continuing to preach this out-worn creed.[39] Before long the group of intellectuals who had once stressed the relative unimportance of the 'material' weapons of the League began to insist loudly on economic and military sanctions as the necessary cornerstones of an international order. When Germany annexed Austria, Lord Cecil indignantly enquired whether the Prime Minister 'holds that the use of material force is impracticable and that the League should cease to attempt "sanctions" and confine its efforts to moral force'.[40] The answer might well have been that, if Neville Chamberlain did in fact hold this view, he could have learned it from Lord Cecil's own earlier utterances.

Moreover, scepticism attacked not only the premise that public opinion is certain to prevail, but also the premise that public opinion is certain to be right. At the Peace Conference, it had been observed that statesmen were sometimes more reasonable and moderate in their demands than the public opinion which they were supposed to represent. Even Wilson himself once used – no doubt, in perfect sincerity – an argument which directly contradicted his customary thesis that reason can be made to prevail by appealing to 'the plain people everywhere throughout the world'. In the League of Nations Commission of the Conference, the Japanese had raised the issue of race equality. 'How can you treat on its merits in this quiet room', enquired the President, 'a question which will not be treated on its merits when it gets out of this room?'[41] Later history provided many examples of this phenomenon. It became a commonplace for statesmen at Geneva and elsewhere to explain that they themselves had every desire to be reasonable, but that public opinion in their countries was inexorable; and though this plea was sometimes a pretext or a tactical manœuvre, there was often a solid substratum of reality beneath it. The prestige of public opinion correspondingly declined. 'It does not help the conciliator, the arbitrator, the policeman or the judge', wrote a well-known supporter of the League of Nations Union recently, 'to be surrounded by a crowd emitting either angry or exulting cheers.'[42] Woodrow Wilson's 'plain men throughout the world', the spokesmen of 'the common purpose of enlightened mankind', had somehow transformed themselves into a disorderly mob emitting incoherent and unhelpful noises. It seemed undeniable that, in international affairs, public opinion was almost as often wrong-headed as it was impotent. But where so many of the presuppositions of 1919 were crumbling, the intellectual leaders of the utopian school stuck to their guns; and in Great Britain and the United States – and to a lesser degree in France – the rift between theory and practice assumed alarming dimensions. Armchair students of international affairs were unanimous about the kind of policy which ought to be followed, both in the political and in the economic field. Governments of many countries acted in a sense precisely contrary to this advice, and received the endorsement of public opinion at the polls.

The problem of diagnosis

In such disasters the obvious explanation is never far to seek. The able historian of the Communist International has noted that, in the history of that institution, 'every failure – not objective failure, but the failure of the reality to comply with the utopia – supposes a traitor'.[43] The principle has a wide application, and touches deep springs of human character. Statesmen of more than one country have been pilloried by disappointed utopians as wreckers of the international order. The few members of the school who have tried to go behind this simple anthropomorphic explanation hesitate between two alternative diagnoses. If mankind in its international relations has signally failed to achieve the rational good, it must either have been too stupid to understand that good, or too wicked to pursue it. Professor Zimmern leans to the hypothesis of stupidity, repeating almost word for word the argument of Buckle and Sir Norman Angell:

> The obstacle in our path ... is not in the moral sphere, but in the intellectual ... It is not because men are ill-disposed that they cannot be educated into a world social consciousness. It is because they – let us be honest and say 'we' – are beings of conservative temper and limited intelligence.

The attempt to build a world order has failed not through 'pride or ambition or greed', but through 'muddled thinking'.[44] Professor Toynbee, on the other hand, sees the cause of the breakdown in human wickedness. In a single volume of the annual *Survey of International Affairs*, he accuses Italy of 'positive, strong-willed, aggressive egotism', Great Britain and France of 'negative, weak-willed, cowardly egotism', Western Christendom as a whole of a 'sordid' crime, and all the members of the League of Nations, except Abyssinia, of 'covetousness' or 'cowardice' (the choice is left to them), while the attitude of the Americans is merely 'rather captious and perverse'.[45] Some writers combined the charge of stupidity and the charge of wickedness. Much comment on international affairs was rendered tedious and sterile by incessant girding at a reality which refused to conform to utopian prescriptions.

The simplicity of these explanations seemed almost ludicrously disproportionate to the intensity and complexity of the international crisis.[46] The impression made on the ordinary man was more accurately recorded in April 1938 in some words of Mr Anthony Eden:

> It is utterly futile to imagine that we are involved in a European crisis which may pass as it has come. We are involved in a crisis of humanity all the world over. We are living in one of those great periods of history which are awe-inspiring in their responsibilities and in their con- sequences. Stupendous forces are loose, hurricane forces.[47]

It is not true, as Professor Toynbee believes, that we have been living in an exceptionally wicked age. It is not true, as Professor Zimmern implies, that we have been living in an exceptionally stupid one. Still less is it true, as

Professor Lauterpacht more optimistically suggests, that what we have been experiencing is 'a transient period of retrogression' which should not be allowed unduly to colour our thought.[48] It is a meaningless evasion to pretend that we have witnessed, not the failure of the League of Nations, but only the failure of those who refused to make it work. The breakdown of the nineteen-thirties was too overwhelming to be explained merely in terms of individual action or inaction. Its downfall involved the bankruptcy of the postulates on which it was based. The foundations of nineteenth-century belief are themselves under suspicion. It may be not that men stupidly or wickedly failed to apply right principles, but that the principles themselves were false or inapplicable. It may turn out to be untrue that if men reason rightly about international politics they will also act rightly, or that right reasoning about one's own or one's nation's interests is the road to an international paradise. If the assumptions of nineteenth-century liberalism are in fact untenable, it need not surprise us that the utopia of the international theorists made so little impression on reality. But if they are untenable today, we shall also have to explain why they found such widespread acceptance, and inspired such splendid achievements, in the nineteenth century.

Notes

1. While this is the form of utopianism which has been predominant for the past three centuries, and which still prevails (though perhaps with diminishing force) in English-speaking countries, it would be rash to assert that individualism and rationalism are necessary attributes to utopian thought. Fascism contained elements of a utopianism which was anti-individualist and irrational. These qualities were already latent in the utopian aspects of Leninism – and perhaps even of Marxism.
2. Bentham, *Works*, ed. Bowring, i. p. 31.
3. Bentham, *Works*, ed. Bowring, viii. p. 561.
4. James Mill, *The Liberty of the Press*, pp. 22–3.
5. J. S. Bury, *The Idea of Progress*, p. 131.
6. Comte, *Cours de Philosophie Positive*, Lecture LXI.
7. Buckle, *History of Civilisation* (World's Classics ed.), i. pp. 151–2.
8. Quoted in Angell, *Foundations of International Polity*, p. 224. Internal evidence suggests that the passage was written by Sir Norman Angell himself.
9. R. H. S. Crossman in J. P. Mayer, *Political Thought*, p. 202.
10. New Year's Eve broadcast from Radio-Nations, Geneva: *The Times*, 1 January 1938.
11. *The Covenant of the League of Nations and a Commentary Thereon*, Cmd 151 (1919), p. 12. 'The great strength of the Covenant', said the British Government some years later, 'lies in the measure of discretion which it allows to the Council and Assembly in dealing with future contingencies

which may have no parallel in history and which therefore cannot all of them be foreseen in advance' (*League of Nations: Official Journal*, May 1928, p. 703).

12. The defection of the United States upset this balance, and left four major confronted with four minor Powers. Subsequent increases in membership, which have taken place at frequent intervals since 1923, gave a permanent preponderance to the minor Powers. The Council, in becoming more 'representative', lost much of its effectiveness as a political instrument. Reality was sacrificed to an abstract principle. It should be added that the prudent Swiss Delegate foresaw this result when the first increase was mooted in 1922 (*League of Nations: Third Assembly*, First Committee, pp. 37–8).

13. By a curious irony, this development was strongly encouraged by a group of American intellectuals; and some European enthusiasts imagined that, by following this course, they would propitiate American opinion. The rift between the theory of the intellectuals and the practice of the government, which developed in Great Britain from 1932 onwards, began in the United States in 1919.

14. J. Fischer-Williams, *Some Aspects of the Covenant of the League of Nations*, p. 238.

15. Winston Churchill, *Arms and the Covenant*, p. 43.

16. Quoted in Toynbee, *Survey of International Affairs*, 1935, ii, p. 448.

17. B. Russell in *Atlantic Monthly*, clix. (February 1937), p. 155.

18. *League of Nations: Fifth Assembly*, p. 497.

19. W. Taft, *The United States and Peace*, p. 150.

20. *The Public Papers of Woodrow Wilson: The New Democracy*, ed. R. S. Baker, i. p. 206.

21. R. S. Baker, *Woodrow Wilson: Life and Letters*, iii. p. 173.

22. *The Public Papers of Woodrow Wilson: War and Peace*, ed. R. S. Baker, i. p. 259.

23. *Intimate Papers of Colonel House*, ed. C. Seymour, iv. p. 291.

24. D. Lloyd George, *The Truth about the Treaties*, i. p. 187.

25. *The Public Papers of Woodrow Wilson: War and Peace*, ed. R. S. Baker, i. p. 133.

26. *Ibid.*, i. p. 234.

27. Miller, *The Drafting of the Covenant*, ii. p. 64.

28. *The Covenant of the League of Nations with a Commentary Thereon*, Cmd 151, pp. 12, 16.

29. House of Commons, 21 July 1919: *Official Report*, cols. 990, 992.

30. *Imperial Conference of 1923*, Cmd 1987, p. 44.

31. *League of Nations: First Assembly*, p. 395.

32. Quoted in Zimmern, *The League of Nations and the Rule of Law*, p. 399.

33. House of Commons, 22 March 1932: *Official Report*, col. 923.

34. N. M. Butler, *The Path to Peace*, p. xii.

35. *League of Nations: Twelfth Assembly*, p. 59.

36. Mr Stimson to the Council of Foreign Relations on 8 August 1932 (*New York Times*, 9 August 1932).

37. 'Believing, as this government does, in the moral force of public opinion ...' (Sumner Welles in *State Department Press Releases*, 8 October 1938, p. 237).

38. *The Times*, 18 April 1939.

39. Winston Churchill, *Arms and the Covenant*, p. 36.
40. *Daily Telegraph*, 24 March 1938.
41. Miller, *The Drafting of the Covenant*, ii. p. 701.
42. Lord Allen of Hurtwood, *The Times*, 30 May 1938.
43. F. Borkenau, *The Communist International*, p. 179.
44. *Neutrality and Collective Security* (Harris Foundation Lectures: Chicago, 1936), pp. 8, 18.
45. *Survey of International Affairs, 1935*, ii. pp. 2, 89, 96, 219–20, 480.
46. As a recent writer has said of the French eighteenth-century rationalists, 'their superficiality lay in a shocking exaggeration of the simplicity of the problem' (Sabine, *A History of Political Theory*, p. 551).
47. Anthony Eden, *Foreign Affairs*, p. 275.
48. *International Affairs*, xvii. (September–October 1938), p. 712.

CHAPTER FOUR

The Harmony of Interests

The utopian synthesis

No political society, national or international, can exist unless people submit to certain rules of conduct. The problem why people should submit to such rules is the fundamental problem of political philosophy. The problem presents itself just as insistently in a democracy as under other forms of government and in international as in national politics; for such a formula as 'the greatest good of the greatest number' provides no answer to the question why the minority, whose greatest good is *ex hypothesi* not pursued, should submit to rules made in the interest of the greatest number. Broadly speaking, the answers given to the question fall into two categories, corresponding to the antithesis, discussed in a previous chapter, between those who regard politics as a function of ethics and those who regard ethics as a function of politics.

Those who assert the primacy of ethics over politics will hold that it is the duty of the individual to submit for the sake of the community as a whole, sacrificing his own interest to the interest of others who are more numerous, or in some other way more deserving. The 'good' which consists in self-interest should be subordinated to the 'good' which consists in loyalty and self-sacrifice for an end higher than self-interest. The obligation rests on some kind of intuition of what is right and cannot be demonstrated by rational argument. Those, on the other hand, who assert the primacy of politics over ethics, will argue that the ruler rules because he is the stronger, and the ruled submit because they are the weaker. This principle is just as easily applicable to democracy as to any other form of government. The majority rules because it is stronger, the minority submits because it is weaker. Democracy, it has often been said, substitutes the counting of heads for the breaking of heads. But the substitution is merely a convenience, and the principle of the two methods is the same. The realist, therefore, unlike the intuitionist, has a perfectly rational answer to the question why the individual should submit. He should submit because otherwise the stronger will compel him; and the results of compulsion are more disagreeable than those of voluntary submission. Obligation is thus derived from a sort of spurious ethic based on the reasonableness of recognizing that might is right.

Both these answers are open to objection. Modern man, who has witnessed so many magnificent achievements of human reason, is reluctant to believe that reason and obligation sometimes conflict. On the other hand, men of all ages have failed to find satisfaction in the view that the rational basis of obligation is merely the right of the stronger. One of the strongest points of eighteenth- and nineteenth-century utopianism was its apparent success in meeting both these objections at once. The utopian, starting from the primacy of ethics, necessarily believes in an obligation which is ethical in character and independent of the right of the stronger. But he has also been able to convince himself, on grounds other than those of the realist, that the duty of the individual to submit to rules made in the interest of the community can be justified in terms of reason, and that the greatest good of the greatest number is a rational end even for those who are not included in the greatest number. He achieves this synthesis by maintaining that the highest interest of the individual and the highest interest of the community naturally coincide. In pursuing his own interest, the individual pursues that of the community, and in promoting the interest of the community he promotes his own. This is the famous doctrine of the harmony of interests. It is a necessary corollary of the postulate that moral laws can be established by right reasoning. The admission of any ultimate divergence of interests would be fatal to this postulate; and any apparent clash of interests must therefore be explained as the result of wrong calculation. Burke tacitly accepted the doctrine of identity when he defined expediency as 'that which is good for the community and for every individual in it'.[1] It was handed on from the eighteenth-century rationalists to Bentham, and from Bentham to the Victorian moralists. The utilitarian philosophers could justify morality by the argument that, in promoting the good of others, one automatically promotes one's own. Honesty is the best policy. If people or nations behave badly, it must be, as Buckle and Sir Norman Angell and Professor Zimmern think, because they are unintellectual and short-sighted and muddle-headed.

The paradise of *laissez-faire*

It was the *laissez-faire* school of political economy created by Adam Smith which was in the main responsible for popularizing the doctrine of the harmony of interests. The purpose of the school was to promote the removal of state control in economic matters; and in order to justify this policy, it set out to demonstrate that the individual could be relied on, without external control, to promote the interests of the community for the very reason that those interests were identical with his own. This proof was the burden of *The Wealth of Nations*. The community is divided into those who live by rent, those who live by wages and those who live by profit; and the interests of 'those three great orders' are 'strictly and inseparably connected with the general interest of the society'.[2] The harmony is none the less real if those concerned are unconscious of it. The individual 'neither intends to promote the public interest, nor knows how much he is promoting it . . . He intends only his own gain, and he is in this, as in many other cases, led by an invisible hand to promote an end which was no part of his intention.'[3] The invisible hand, which Adam Smith would perhaps have regarded as a metaphor, presented no difficulty to Victorian piety. 'It is curious to observe', remarks a tract issued by the Society for the Propagation of Christian Knowledge towards the middle of the nineteenth century, 'how, through the wise and beneficent arrangement of Providence, men thus do the greatest service to the public when they are thinking of nothing but their own gain.'[4] About the same time an English clergyman wrote a work entitled *The Temporal Benefits of Christianity Explained*. The harmony of interests provided a solid rational basis for morality. To love one's neighbour turned out to be a thoroughly enlightened way of loving oneself. 'We now know', wrote Mr Henry Ford as recently as 1930, 'that anything which is economically right is also morally right. There can be no conflict between good economics and good morals.'[5]

The assumption of a general and fundamental harmony of interests is *prima facie* so paradoxical that it requires careful scrutiny. In the form which Adam Smith gave to it, it had a definite application to the economic structure of the eighteenth century. It presupposed a society of small producers and merchants, interested in the maximization of production and exchange, infinitely mobile and adaptable, and unconcerned with the problem of the distribution of wealth. Those conditions were substantially fulfilled in an age when production involved no high degree of specialization and no sinking of capital in fixed equipment, and when the class which might be more interested in the equitable distribution of wealth than in its maximum production was insignificant and without influence. But by a curious coincidence, the year which saw the publication of *The Wealth of Nations* was also the year in which Watt invented his steam engine. Thus, at the very moment when *laissez-faire* theory was receiving its classical exposition, its premises were undermined by an invention which was destined to call into being immobile, highly specialized, mammoth industries and a large and powerful proletariat more interested in distribution than in production. Once industrial capitalism and the class system had become the recognized structure of society, the doctrine of the harmony of interests acquired a new significance, and became, as we shall presently see, the ideology of a dominant group concerned to maintain its predominance by asserting the identity of its interests with those of the community as a whole.[6]

But this transformation could not have been effected, and the doctrine could not have survived at all, but for one circumstance. The survival of the belief in a harmony of interests was rendered possible by the unparalleled expansion of production, population and prosperity, which marked the hundred years following the publication of *The Wealth of Nations* and the invention of the steam engine. Expanding prosperity contributed to the popularity of the doctrine in three different ways. It attenuated competition for markets among producers, since fresh markets were constantly available; it postponed the class issue, with its insistence on the primary importance of equitable distribution, by extending to members of the less prosperous classes some share in the general prosperity; and by creating a sense of confidence in present and future well-being, it encouraged men to believe that the world was ordered on so rational a plan as the natural harmony of interests. 'It was the continual widening of the field of demand which, for half a century, made capitalism operate as if it were a liberal utopia.'[7] The tacit presupposition of infinitely expanding markets was the foundation on which the supposed harmony of interests rested. As Dr Mannheim points out, traffic control is unnecessary so long as the number of cars does not exceed the comfortable capacity of the road.[8] Until that moment arrives, it is easy to believe in a natural harmony of interests between road-users.

What was true of individuals was assumed to be also true of nations. Just as individuals, by pursuing their own good, unconsciously compass the good of the whole community, so nations in serving themselves serve humanity. Universal free trade was justified on the ground that the maximum economic interest of each nation was identified with the maximum economic interest of the whole world. Adam Smith, who was a practical reformer rather than a pure theorist, did indeed admit that governments might have to protect certain industries in the interests of national defence. But such derogations seemed to him and to his followers trivial exceptions to the rule. '*Laissez-faire*', as J. S. Mill puts it, '... should be the general rule: every departure from it, unless required by some great good, a certain evil.'[9] Other thinkers gave the doctrine of the harmony of national interests a still wider application. 'The true interests of a nation', observes a late eighteenth-century writer, 'never yet stood in opposition to the general interest of mankind; and it can never happen that philanthropy and patriotism can impose on any man inconsistent duties.'[10] T. H. Green, the English Hegelian who tempered the doctrines of his master with concessions to British nineteenth-century liberalism, held that 'no action in its own interest of a state which fulfilled its idea could conflict with any true interest or right of general society',[11] though it is interesting to note that the question-begging epithet 'true', which in the eighteenth-century quotation is attached to the interests of the nation, has been transferred by the nineteenth century to the interest of the general society. Mazzini, who embodied the liberal nineteenth-century philosophy of nationalism, believed in a sort of division of labour between nations. Each nation had its own special task for which its special aptitudes fitted it, and the performance of this task was its contribution to the welfare of humanity. If all nations acted in this spirit, international harmony would prevail. The same condition of apparently infinite expansibility which encouraged belief in the economic harmony of interests made possible the belief in the political harmony of rival national movements. One reason why contemporaries of Mazzini thought nationalism a good thing was that there were few recognized nations, and plenty of room for them. In an age when Germans, Czechs, Poles, Ukrainians, Magyars and half a dozen more national groups were not yet visibly jostling one another over an area of a few hundred square miles, it was comparatively easy to believe that each nation, by developing its own nationalism, could make its own special contribution to the international harmony of interests. Most liberal writers continued to believe, right down to 1918, that nations, by developing their own nationalism, promoted the cause of internationalism; and Wilson and many other makers of the peace treaties saw in national self-determination the key to world peace. More recently still, responsible Anglo-Saxon statesmen have been from time to time content to echo, probably without much reflexion, the old Mazzinian formulae.[12]

Darwinism in politics

When the centenary of *The Wealth of Nations* was celebrated in 1876, there were already symptoms of an impending breakdown. No country but Great Britain had been commercially powerful enough to believe in the international harmony of economic interests. Acceptance of free-trade principles outside Great Britain had always been partial, half-hearted and short-lived. The United States had rejected them from the start. About 1840, Friedrich List, who had spent much time studying industrial development in the United States, began to preach to a German audience the doctrine that, while free trade was the right policy for an industrially dominant nation like Great Britain, only protection could enable weaker nations to break the British stranglehold. German and American industries, built up behind protective tariffs, were soon seriously impinging on the worldwide British industrial monopoly. The British Dominions overseas made use of their newly-won fiscal autonomy to protect themselves against the manufactures of the mother country. The pressure of competition was increasing on all sides. Nationalism began to wear a sinister aspect, and to degenerate into imperialism. The philosophy of Hegel, who identified reality with an eternally recurring conflict of ideas, extended its influence. Behind Hegel stood Marx, who materialized the Hegelian conflict into a class-war of economic interest-groups, and working-class parties came into being which steadfastly refused to believe in the harmony of interests between capital and labour. Above all, Darwin propounded and popularized a biological doctrine of evolution through a perpetual struggle for life and the elimination of the unfit.

It was the doctrine of evolution which for a time enabled the *laissez-faire* philosophy to make its terms with the new conditions and the new trend of thought. Free competition had always been worshipped as the beneficent deity of the *laissez-faire* system. The French economist Bastiat, in a work significantly entitled *Les Harmonies Économiques*, had hailed competition as 'that humanitarian force ... which continually wrests progress from the hands of the individual to make it the common heritage of the great human family'.[13] Under the growing strains of the latter half of the nineteenth century, it was perceived that competition in the economic sphere implied exactly what Darwin proclaimed as the biological law of nature – the survival of the stronger at the expense of the weaker. The small producer or trader was gradually being put out of business by his large-scale competitor; and this development was what progress and the welfare of the community as a whole demanded. *Laissez-faire* meant an open field, and the prize to the strongest. The doctrine of the harmony of interests underwent an almost imperceptible modification. The good of the community (or, as people were now inclined to say, of the species) was still identical with the good of its individual members, but only of those individuals who were effective competitors in the struggle for life. Humanity went on from strength to strength, shedding its weaklings by the way. 'The development of the species,' as Marx said, '... and therefore the higher development of the individual, can only be secured through the historical process, in which individuals are sacrificed.'[14] Such was the doctrine of the new age of intensified economic competition preached by the school of Herbert Spencer, and commonly accepted in Great Britain in the seventies and eighties. The last French disciple of Adam Smith, Yves Guyot, assisted perhaps by the accident that the French word *concurrence* means 'collaboration' as well as 'competition', wrote a work entitled *La Morale de la Concurrence*. Among English writers who applied this evolutionary principle to international politics, the most popular was Bagehot:

> Conquest is the premium given by nature to those national characters which their national customs have made most fit to win in war, and in most material respects those winning characters are really the best characters. The characters which do win in war are the characters which we should wish to win in war.[15]

About the same time, a Russian sociologist defined international politics as 'the art of conducting the struggle for existence between social organisms';[16] and in 1900 a distinguished professor, in a once famous book, stated the doctrine in all its naked ruthlessness:

> The path of progress is strewn with the wreck of nations; traces are everywhere to be seen of the hecatombs of inferior races, and of victims who found not the narrow way to the greater perfection. Yet these dead peoples are, in very truth, the stepping stones on which mankind has arisen to the higher intellectual and deeper emotional life of to-day.[17]

In Germany, the same view was propounded by Treitschke and Houston Stewart Chamberlain. The doctrine of progress through the elimination of unfit nations seemed a fair corollary of the doctrine of progress through the elimination of unfit individuals; and some such belief, though not always openly avowed, was implicit in late nineteenth-century imperialism. In the later nineteenth century, as an American historian remarks, 'the basic problem of international relations was who should cut up the victims'.[18] The harmony of interests was established through the sacrifice of 'unfit' Africans and Asiatics.

One point had, unfortunately, been overlooked. For more than a hundred years, the doctrine of the harmony of interests had provided a rational basis for morality. The individual had been urged to serve the interest of the community on the plea that that interest was also his own. The ground had now been shifted. In the long run, the good of the community and the good of the individual were still the same. But this eventual harmony was preceded by a struggle for life between individuals, in which not only the good, but the very existence, of the loser were eliminated altogether from the picture. Morality in these conditions had no rational attraction for prospective losers; and the whole ethical system was built on the sacrifice of the weaker brother. In practice, nearly every state had made inroads on the classical doctrine, and introduced social legislation to protect the economically weak against the economically strong. The doctrine itself died harder. In the seventies Dostoevsky, who had none of the prejudices of an Englishman or an economist, made Ivan Karamazov declare that the price of admission to the 'eternal harmony' was too high if it included the sufferings of the innocent. About the same time, Winwood Reade made an uncomfortable sensation in Great Britain with a book called *The Martyrdom of Man*, which drew attention to the immense tale of suffering and waste involved in the theory of evolution. In the nineties, Huxley confessed, in the name of science, to the existence of a discrepancy between the 'cosmic process' and the 'ethical process';[19] and Balfour, approaching the problem from the angle of philosophy, concluded that 'a complete harmony between "egoism" and "altruism", between the pursuit of the highest happiness for oneself and the highest happiness for other people, can never be provided by a creed which refuses to admit that the deeds done and the character formed in this life can flow over into another, and there permit a reconciliation and an adjustment between the conflicting principles which are not always possible here'.[20] Less and less was heard of the beneficent properties of free competition. Before 1914, though the policy of international free trade was still upheld by the British electorate and by British economists, the ethical postulate which had once formed the basis of the *laissez-faire* philosophy no longer appealed, at any rate in its crude form, to any serious thinker. Biologically and economically, the doctrine of the harmony of interests was tenable only if you left out of account the interest of the weak who must be driven to the wall, or called in the next world to redress the balance of the present.

The international harmony

Attention has been drawn to the curious way in which doctrines, already obsolete or obsolescent before the war of 1914, were reintroduced in the post-war period, largely through American inspiration, into the special field of international affairs. This would appear to be conspicuously true of the *laissez-faire* doctrine of the harmony of interests. In the United States, the history of *laissez-faire* presents special features. Throughout the nineteenth, and well into the twentieth, centuries the United States, while requiring tariff protection against European competition, had enjoyed the advantage of an expanding domestic market of apparently unlimited potentialities. In Great Britain, which continued down to 1914 to dominate world trade, but was increasingly conscious of strains and stresses at home, J. S. Mill and later economists clung firmly to international free trade, but made more and more inroads into *laissez-faire* orthodoxy in the domestic sphere. In the United States, Carey and his successors justified protective tariffs, but in every other respect maintained the immutable principles of *laissez-faire*. In Europe after 1919, planned economy, which rests on the assumption that no natural harmony of interests exists and that interests must be artificially harmonized by state action, became the practice, if not the theory, of almost every state. In the United States, the persistence of an expanding domestic market staved off this development till after 1929. The natural harmony of interests remained an integral part of the American view of life; and in this, as in other respects, current theories of international politics were deeply imbued with the American tradition. Moreover, there was a special reason for the ready acceptance of the doctrine in the international sphere. In domestic affairs it is clearly the business of the state to create harmony if no natural harmony exists. In international politics, there is no organized power charged with the task of creating harmony; and the temptation to assume a natural harmony is therefore particularly strong. But this is no excuse for burking the issue. To make the harmonization of interests the goal of political action is not the same thing as to postulate that a natural harmony of interests exists;[21] and it is this latter postulate which has caused so much confusion in international thinking.

The common interest in peace

Politically, the doctrine of the identity of interests has commonly taken the form of an assumption that every nation has an identical interest in peace, and that any nation which desires to disturb the peace is therefore both irrational and immoral. This view bears clear marks of its Anglo-Saxon origin. It was easy after 1918 to convince that part of mankind which lives in English-speaking countries that war profits nobody. The argument did not seem particularly convincing to Germans, who had profited greatly from the wars of 1866 and 1870, and attributed their more recent sufferings, not to the war of 1914, but to the fact that they had lost it; or to Italians, who blamed not the war, but the treachery of allies who defrauded them in the peace settlement; or to Poles or Czecho-Slovaks who, far from deploring the war, owed their national existence to it; or to Frenchmen, who could not unreservedly regret a war which had restored Alsace-Lorraine to France; or to people of other nationalities who remembered profitable wars waged by Great Britain and the United States in the past. But these people had fortunately little influence over the formation of current theories of international relations, which emanated almost exclusively from the English-speaking countries. British and American writers continued to assume that the uselessness of war had been irrefutably demonstrated by the experience of 1914–18, and that an intellectual grasp of this fact was all that was necessary to induce the nations to keep the peace in the future; and they were sincerely puzzled as well as disappointed at the failure of other countries to share this view.

The confusion was increased by the ostentatious readiness of other countries to flatter the Anglo-Saxon world by repeating its slogans. In the fifteen years after the first world war, every Great Power (except, perhaps, Italy) repeatedly did lip-service to the doctrine by declaring peace to be one of the main objects of its policy.[22] But as Lenin observed long ago, peace in itself is a meaningless aim. 'Absolutely everybody is in favour of peace in general,' he wrote in 1915, 'including Kitchener, Joffre, Hindenburg and Nicholas the Bloody, for everyone of them wishes to end the war.'[23] The common interest in peace masks the fact that some nations desire to maintain the *status quo* without having to fight for it, and others to change the *status quo* without having to fight in order to do so.[24] The statement that it is in the interest of the world as a whole either that the *status quo* should be maintained, or that it should be changed, would be contrary to the facts. The statement that it is in the interest of the world as a whole that the conclusion eventually reached, whether maintenance or change, should be reached by peaceful means, would command general assent, but seems a rather meaningless platitude. The utopian assumption that there is a world interest in peace which is identifiable with the interest of each individual nation helped politicians and political writers everywhere to evade the unpalatable fact of a fundamental divergence of interest between nations desirous of maintaining the *status quo* and nations desirous of changing it.[25] A peculiar combination of platitude and falseness thus became endemic in the pronouncements of statesmen about international affairs. 'In this whole Danubian area,' said a Prime Minister of Czecho-Slovakia, 'no one really wants conflicts and jealousies. The various countries want to maintain their independence, but otherwise they are ready for any co-operative measures. I am thinking specially of the Little Entente, Hungary and Bulgaria.'[26] Literally the words may pass as true. Yet the conflicts and jealousies which nobody wanted were a notorious feature of Danubian politics after 1919, and the co-operation for which all were ready was unobtainable. The fact of divergent interests was disguised and falsified by the platitude of a general desire to avoid conflict.

International economic harmony

In economic relations, the assumption of a general harmony of interests was made with even greater confidence; for here we have a direct reflexion of the cardinal doctrine of *laissez-faire* economics, and it is here that we can see most clearly the dilemma which results from the doctrine. When the nineteenth-century liberal spoke of the greatest good of the greatest number, he tacitly assumed that the good of the minority might have to be sacrificed to it. This principle applied equally to international economic relations. If Russia or Italy, for example, were not strong enough to build up industries without the protection of tariffs, then – the *laissez-faire* liberal would have argued – they should be content to import British and German manufactures and supply wheat and oranges to the British and German markets. If anyone had thereupon objected that this policy would condemn Russia and Italy to remain second-rate Powers economically and militarily dependent on their neighbours, the *laissez-faire* liberal would have had to answer that this was the will of Providence and that this was what the general harmony of interests demanded. The modern utopian internationalist enjoys none of the advantages, and has none of the toughness, of the nineteenth-century liberal. The material success of the weaker Powers in building up protected industries, as well as the new spirit of internationalism, preclude him from arguing that the harmony of interests depends on the sacrifice of economically unfit nations. Yet the abandonment of this premiss destroys the whole basis of the doctrine which he has inherited; and he is driven to the belief that the common good can be achieved without any sacrifice of the good of any individual member of the community. Every international conflict is therefore unnecessary and illusory. It is only necessary to discover the common good which is at the same time the highest good of all the disputants; and only the folly of statesmen stands in the way of its discovery. The utopian, secure in his understanding of this common good, arrogates to himself the monopoly of wisdom. The statesmen of the world one and all stand convicted of incredible blindness to the interest of those whom they are supposed to represent. Such was the picture of the international scene presented, in all seriousness, by British and American writers, including not a few economists.

It is for this reason that we find in the modern period an extraordinary divergence between the theories of economic experts and the practice of those responsible for the economic policies of their respective countries. Analysis will show that this divergence springs from a simple fact. The economic expert, dominated in the main by *laissez-faire* doctrine, considers the hypothetical economic interest of the world as a whole, and is content to assume that this is identical with the interest of each individual country. The politician pursues the concrete interest of his country, and assumes (if he makes any assumption at all) that the interest of the world as a whole is identical with it. Nearly every pronouncement of every international economic conference held between the two world wars was vitiated by this assumption that there was some 'solution' or 'plan' which, by a judicious balancing of interests, would be equally favourable to all and prejudicial to none.

Any strictly nationalistic policy [declared the League Conference of economic experts in 1927] is harmful not only to the nation which practises it but also to the others, and therefore defeats its own end, and if it be desired that the new state of mind revealed by the Conference should lead rapidly to practical results, any programme of execution must include, as an essential factor, the principle of *parallel* or *concerted* action by the different nations. Every country will then know that the concessions it is asked to make will be balanced by corresponding sacrifices on the part of the other countries. It will be able to accept the proposed measures, not merely in view of its own individual position, *but also because it is interested in the success of the general plan laid down by the Conference.*[27]

The sequel of the Conference was the complete neglect of all the recommendations unanimously made by it; and if we are not content to accept the facile explanation that the leading statesmen of the world were either criminal or mad, we may begin to suspect the validity of its initial assumption. It seems altogether rash to suppose that economic nationalism is necessarily detrimental to states which practise it. In the nineteenth century, Germany and the United States, by pursuing a 'strictly nationalistic policy', had placed themselves in a position to challenge Great Britain's virtual monopoly of world trade. No conference of economic experts, meeting in 1880, could have evolved a 'general plan' for 'parallel or concerted action' which would have allayed the economic rivalries of the time in a manner equally advantageous to Great Britain, Germany and the United States. It was not less presumptuous to suppose that a conference meeting in 1927 could allay the economic rivalries of the later period by a 'plan' beneficial to the interests of everyone. Even the economic crisis of 1930–33 failed to bring home to the economists the true nature of the problem which they had to face. The experts who prepared the 'Draft Annotated Agenda' for the World Economic Conference of 1933 condemned the 'worldwide adoption of ideals of national self-sufficiency which cut unmistakably athwart the lines of economic development'.[28] They did not apparently pause to reflect that those so-called 'lines of economic development', which might be beneficial to some countries and even to the world as a whole, would inevitably be detrimental to other countries, which were using weapons of economic nationalism in self-defence. The Van Zeeland report of January 1938 began by asking, and answering in the affirmative, the question whether 'the methods which, taken as a whole, form the system of international trade' are 'fundamentally preferable' to 'autarkic tendencies'. Yet every Power at some period of its history, and as a rule for prolonged periods, has resorted to 'autarkic tendencies'. It is difficult to believe that there is any absolute sense in which 'autarkic tendencies' are always detrimental to those who pursue them. Even if they could be justified only as the lesser of two evils, the initial premise of the Van Zeeland report was invalidated. But there was worse to come. 'We must ... make our dispositions', continued M. Van Zeeland, 'in such a way that the new system shall offer to all participators advantages greater than those of the position in which they now find themselves.'[29] This is economic utopianism in its most purblind form. The report, like the reports of 1927 and 1933, assumed the existence of a fundamental principle of economic policy whose application would be equally beneficial to all states and detrimental to none; and for this reason it remained, like its predecessors, a dead letter.

Economic theory, as opposed to economic practice, was so powerfully dominated in the years between the two world wars by the supposed harmony of interests that it is difficult to find, in the innumerable international discussions of the period, any clear exposition of the real problem which baffled the statesmen of the world. Perhaps the frankest statement was one made by the Yugoslav Foreign Minister at the session of the Commission for European Union in January 1931. Arthur Henderson, on behalf of Great Britain, following the Netherlands delegate Dr Colijn, had pleaded for an all-round tariff reduction 'which must, by its nature, bring benefit to each and all by allowing that expansion of production and international exchange of wealth by which the common prosperity of all can be increased'.[30] Marinkovitch, who spoke next, concluded from the failure to carry out the recommendations of the 1927 Conference, that 'there were extremely important reasons why the governments could not apply' those resolutions. He went on:

The fact is that apart from economic considerations there are also political and social considerations. The old 'things-will-right-themselves' school of economists argued that if nothing were done and events were allowed to follow their natural course from an economic point of view, economic equilibrium would come about of its own accord. That is probably true (I do not propose to discuss the point). But how would that equilibrium come about? At the expense of the weakest. Now, as you are aware, for more than seventy years there has been a powerful and growing reaction against this theory of economics. All the socialist parties of Europe and the world are merely the expression of the opposition to this way of looking at economic problems.

We were told that we ought to lower customs barriers and even abolish them. As far as the agricultural states of Europe are concerned, if they could keep the promises they made in 1927 – admitting that the statements of 1927 did contain promises – and could carry that policy right through, we might perhaps find ourselves able to hold our own against overseas competition in the matter of agricultural products. But at the same time we should have to create in Poland, Roumania and Yugoslavia the same conditions as exist in Canada and the Argentine, where vast territories are inhabited by a scanty population and where machinery and other devices are employed. ... We could not sacrifice our people by shooting them, but they would nevertheless be killed off by famine – which would come to the same thing.

I am sure that the key to which M. Colijn has referred does not exist. Economic and social life is too complicated to allow of a solution by any one formula; it calls for complicated solutions. We shall have to take into account the many varieties of geographical, political, social and other conditions which exist.[31]

Marinkovitch went on to dispose of the theory of the 'long-run' harmony of interests:

Last year, when I was in the Yugoslav mountains, I heard that the inhabitants of a small mountain village, having no maize or wheat on which to live, were simply cutting down a wood which belonged to them ... and were living on what they earned by selling the wood. ... I went to the village, collected together some of the leading inhabitants and endeavoured to reason with them, just like the great industrial states reason with us. I said to them: 'You possess plenty of common sense. You see that your forest is becoming smaller and smaller. What will you do when you cut down the last tree?' They replied to me: 'Your Excellency, that is a point which worries us: but on the other hand, what should we do now if we stopped cutting down our trees?'

I can assure you that the agricultural countries are in exactly the same situation. You threaten them with future disasters; but they are already in the throes of disaster.[32]

One further example of unwonted frankness may be quoted. Speaking in September 1937 over one of the United States broadcasting systems, the President of the Colombian Republic said:

In no field of human activity are the benefits of the crisis as clear as in the relationships between nations and especially of the American nations. If it is true that the economic relations have become rigorous and at times harsh, it is also true that they have fortunately become more democratic.

The crisis freed many countries which had up to then been subordinated to the double mental and financial imperialism of the nations which controlled international markets and policies. Many nations learned to trust less international cordiality and to seek an autonomous life, full of initial obstacles but which nevertheless created strong interests within a short time...

When the arbitrary systems that prevail today begin to be relaxed, there will be a weaker international trade, but there will also be a larger number of nations economically strong.

Economic co-operation today is a very different and more noble thing than the old co-operation which was based on the convenience of industrial countries and of bankers who tutored the world. The certainty acquired by many small nations that they can subsist and prosper without subordinating their conduct and their activities to foreign interests has begun to introduce a greater frankness and equality in the relations between modern nations...

It is true that the crisis has shipwrecked many high and noble principles of our civilization; but it is also true that in this return to a kind of primitive struggle for existence, peoples are being freed of many fictions and of much hypocrisy which they had accepted in the belief that with them they were insuring their well-being...

The foundation of international economic freedom lies in the recognition that when strong nations place themselves on the defensive, they act just like the weak ones do, and that all of them have an equal right to defend themselves with their own resources.[33]

The claims made on behalf of the Colombian Republic were perhaps exaggerated. But both the Yugoslav and the Colombian statements were powerful challenges to the doctrine of the harmony of interests. It is fallacy to suppose that, because Great Britain and the United States have an interest in the removal of trade barriers, this is also an interest of Yugoslavia and Colombia. International trade may be weaker. The economic interests of Europe or of the world at large may suffer. But Yugoslavia and Colombia will be better off than they would have been under a régime of European or world prosperity which reduced them to the position of satellites. Dr Schacht spoke a little later of those 'fanatical adherents of the most-favoured-nation policy abroad, who from the abundance of their wealth cannot realize that a poor nation has nevertheless the courage to live by its own laws instead of suffering under the prescriptions of the well-to-do'.[34]

Laissez-faire, in international relations as in those between capital and labour, is the paradise of the economically strong. State control, whether in the form of protective legislation or of protective tariffs, is the weapon of self-defence invoked by the economically weak. The clash of interests is real and inevitable; and the whole nature of the problem is distorted by an attempt to disguise it.

The harmony broken

We must therefore reject as inadequate and misleading the attempt to base international morality on an alleged harmony of interests which identifies the interest of the whole community of nations with the interest of each individual member of it. In the nineteenth century, this attempt met with widespread success, thanks to the continuously expanding economy in which it was made. The period was one of progressive prosperity, punctuated only by minor set-backs. The international economic structure bore considerable resemblance to the domestic economic structure of the United States. Pressure could at once be relieved by expansion to hitherto unoccupied and unexploited territories; and there was a plentiful supply of cheap labour, and of backward countries, which had not yet reached the level of political consciousness. Enterprising individuals could solve the economic problem by migration, enterprising nations by colonization. Expanding markets produced an expanding population, and population in turn reacted on markets. Those who were left behind in the race could plausibly be regarded as the unfit. A harmony of interests among the fit, based on individual enterprise and free competition, was sufficiently near to reality to form a sound basis for the current theory. With some difficulty the illusion was kept alive till 1914. Even British prosperity, though its foundations were menaced by German and American competition, continued to expand. The year 1913 was a record year for British trade.

The transition from the apparent harmony to the transparent clash of interests may be placed about the turn of the century. Appropriately enough, it found its first expression in colonial policies. In the British mind, it was primarily associated with events in South Africa. Mr Churchill dates the beginning of 'these violent times' from the Jameson Raid.[35] In North Africa and the Far East, there was a hasty scramble by the European Powers to secure the few eligible sites which were still vacant. Emigration of individuals from Europe, the point of principal tension, to America assumed unparalleled dimensions. In Europe itself, anti-Semitism – the recurrent symptom of economic stress – reappeared after a long interval in Russia, Germany and France.[36] In Great Britain, agitation against unrestricted alien immigration began in the 1890s; and the first act controlling immigration was passed in 1905.

The first world war, which proceeded from this growing tension, aggravated it tenfold by intensifying its fundamental causes. In belligerent and neutral countries in Europe, Asia and America industrial and agricultural production were everywhere artificially stimulated. After the war every country struggled to maintain its expanded production; and an enhanced and inflamed national consciousness was invoked to justify the struggle. One reason for the unprecedented vindictiveness of the peace treaties, and in particular of their economic clauses, was that practical men no longer believed – as they had done fifty or a hundred years earlier – in an underlying harmony of interests between victors and defeated. The object was now to eliminate a competitor, a revival of whose prosperity might menace your own. In Europe, the struggle was intensified by the creation of new states and new economic frontiers. In Asia, India and China built up large-scale manufactures to make themselves independent of imports from Europe. Japan became an exporter of textiles and other cheap goods which undercut European manufactures on the world market. Most important of all, there were no more open spaces anywhere awaiting cheap and profitable development and exploitation. The ample avenues of migration which had relieved the economic pressures of the pre-war period were closed; and in place of the natural flow of migration came the problem of forcibly evicted refugees.[37] The complex phenomenon known as economic nationalism swept over the world. The fundamental character of this clash of interests became obvious to all except those confirmed utopians who dominated economic thought in the English-speaking countries. The hollowness of the glib nineteenth-century platitude that nobody can benefit from what harms another was revealed. The basic presupposition of utopianism had broken down.

What confronts us in international politics today is, therefore, nothing less than the complete bankruptcy of the conception of morality which has dominated political and economic thought for a century and a half. Internationally, it is no longer possible to deduce virtue from right reasoning, because it is no longer seriously possible to believe that every state, by pursuing the greatest good of the whole world, is pursuing the greatest good of its own citizens, and *vice versa*. The synthesis of morality and reason, at any rate in the crude form in which it was achieved by nineteenth-century liberalism, is untenable. The inner meaning of the modern international crisis is the collapse of the whole structure of utopianism based on the concept of the harmony of interests. The present generation will have to rebuild from the foundations. But before we can do this, before we can ascertain what can be salved from the ruins, we must examine the flaws in the structure which led to its collapse; and we can best do this by analysing the realist critique of the utopian assumptions.

Notes

1. Burke, *Works*, v. 407.
2. Adam Smith, *The Wealth of Nations*, Book I. ch. xi. conclusion.
3. *Ibid.*, Book IV. ch. ii.
4. Quoted in J. M. Keynes, *A Tract on Monetary Reform*, p. 7.
5. Quoted in J. Truslow Adams, *The Epic of America*, p. 400. I have failed to trace the original.
6. See pp. 80–1.
7. *Nationalism: A Study by a Group of Members of the Royal Institute of International Affairs*, p. 229.
8. K. Mannheim, *Mensch und Gesellschaft im Zeitalter des Umbaus*, p. 104.
9. J. S. Mill, *Principles of Political Economy*, II. Book V. ch. xi.
10. Romilly, *Thoughts on the Influence of the French Revolution*, p. 5.
11. T. H. Green, *Principles of Political Obligation*, § 166.
12. Mr Eden, for example, in 1938 advocated 'a comity of nations in which each can develop and flourish and give to their uttermost their own special contribution to the diversity of life' (Anthony Eden, *Foreign Affairs*, p. 277).
13. Bastiat, *Les Harmonies Économiques*, p. 355.
14. Marx, *Theorien über den Mehrwert*, II, i. p. 309.
15. Bagehot, *Physics and Politics* (2nd ed.), p. 215. What does 'material' mean in this passage? Does it merely mean 'relevant'? Or is the writer conscious of an uncomfortable antithesis between 'material' and 'moral'?
16. J. Novicow, *La Politique Internationale*, p. 242.
17. Karl Pearson, *National Life from the Standpoint of Science*, p. 64.
18. W. L. Langer, *The Diplomacy of Imperialism*, ii. p. 797.
19. Huxley, Romanes Lecture, 1893, reprinted in *Evolution and Ethics*, p. 81.
20. Balfour, *Foundations of Belief*, p. 27.
21. The confusion between the two was admirably illustrated by an interjection of Mr Attlee in the House of Commons: 'It was precisely the object of the establishment of the League of Nations that the preservation of peace was a common interest of the world' (House of Commons, 21 December 1937: *Official Report*, col. 1811). Mr Attlee apparently failed to distinguish between the proposition that a natural community of interests existed and the proposition that the League of Nations had been established to create one.
22. 'Peace must prevail, must come before all' (Briand, *League of Nations: Ninth Assembly*, p. 83). 'The maintenance of peace is the first objective of British foreign policy' (Eden, *League of Nations: Sixteenth Assembly*, p. 106). 'Peace is our dearest treasure' (Hitler, in a speech in the German Reichstag on 30

January 1937, reported in *The Times*, 1 February 1937). 'The principal aim of the international policy of the Soviet Union is the preservation of peace' (Chicherin in *The Soviet Union and Peace* (1929), p. 249). 'The object of Japan, despite propaganda to the contrary, is peace' (Matsuoka, *League of Nations: Special Assembly 1932–33*, iii. p. 73). The paucity of Italian pronouncements in favour of peace was probably explained by the poor reputation of Italian troops as fighters: Mussolini feared that any emphatic expression of preference for peace would be construed as an admission that Italy had no stomach for war.

23. Lenin, *Collected Works* (Engl. transl.), xviii. p. 264. Compare Spencer Wilkinson's dictum: 'It is not peace but preponderance that is in each case the real object. The truth cannot be too often repeated that peace is never the object of policy: you cannot define peace except by reference to war, which is a means and never an end' (*Government and the War*, p. 121).

24. 'When a saint complains that people do not know the things belonging to their peace, what he really means is that they do not sufficiently care about the things belonging to his peace' (*The Note-Books of Samuel Butler*, ed. Festing-Jones, pp. 211–12). This would seem to be true of those latter-day saints, the satisfied Powers.

25. It is sometimes maintained not merely that all nations have an equal interest in preferring peace to war (which is, in a sense, true), but that war can never in any circumstances bring to the victor advantages comparable with its cost. The latter view does not appear to be true of the past, though it is possible to argue (as does Bertrand Russell, *Which Way Peace?*) that it is true of modern warfare. If accepted, this view leads, of course, to absolute pacifism; for there is no reason to suppose that it is any truer of 'defensive' than of 'offensive' war (assuming the distinction between them to be valid).

26. *Daily Telegraph*, 26 August 1938.

27. *League of Nations*: C.E.I. 44, p. 21 (italics in original).

28. *League of Nations*: C.48, M.18, 1933, ii. p. 6.

29. *Report ... on the Possibility of Obtaining a General Reduction of the Obstacles to International Trade*, Cmd 5648.

30. *League of Nations*: C.144, M.45, 1931, vii. p. 30.

31. *League of Nations*: C.144, M.45, 1931, vii. p. 31.

32. *Ibid.*, p. 32.

33. Address broadcast by the Columbia Broadcasting System, USA, on 19 September 1937, and published in *Talks*, October 1937.

34. Address to the Economic Council of the German Academy, 29 November 1938.

35. Winston Churchill, *World Crisis*, p. 26.

36. The same conditions encouraged the growth of Zionism; for Zionism, as the Palestine Royal Commission of 1937 remarked, 'on its negative side is a creed of escape' (Cmd 5479, p. 13).

37. 'The existence of refugees is a symptom of the disappearance of economic and political liberalism. Refugees are the by-product of an economic isolationism which has practically stopped free migration' (J. Hope Simpson, *Refugees: Preliminary Report of a Survey*, p. 193).

CHAPTER FIVE

The Realist Critique

The foundations of realism

For reasons explained in a previous chapter, realism enters the field far behind utopianism and by way of reaction from it. The thesis that 'justice is the right of the stranger' was, indeed, familiar in the Hellenic world. But it never represented anything more than the protest of an uninfluential minority, puzzled by the divergence between political theory and political practice. Under the supremacy of the Roman Empire, and later of the Catholic Church, the problem could hardly arise; for the political good, first of the empire, then of the church, could be regarded as identical with moral good. It was only with the break-up of the mediaeval system that the divergence between political theory and political practice became acute and challenging. Machiavelli is the first important political realist.

Machiavelli's starting-point is a revolt against the utopianism of current political thought:

> It being my intention to write a thing which shall be useful to him who apprehends it, it appears to me more appropriate to follow up the real truth of a matter than the imagination of it; for many have pictured republics and principalities which in fact have never been seen and known, because how one lives is so far distant from how one ought to live that he who neglects what is done for what ought to be done sooner effects his ruin than his preservation.

The three essential tenets implicit in Machiavelli's doctrine are the foundation-stones of the realist philosophy. In the first place, history is a sequence of cause and effect, whose course can be analysed and understood by intellectual effort, but not (as the utopians believe) directed by 'imagination'. Secondly, theory does not (as the utopians assume) create practice, but practice theory. In Machiavelli's words, 'good counsels, whencesoever they come, are born of the wisdom of the prince, and not the wisdom of the prince from good counsels'. Thirdly, politics are not (as the utopians pretend) a function of ethics, but ethics of politics. Men 'are kept honest by constraint'. Machiavelli recognized the importance of morality, but thought that there could be no effective morality where there was no effective authority. Morality is the product of power.[1]

The extraordinary vigour and vitality of Machiavelli's challenge to orthodoxy may be attested by the fact that, more than four centuries after he wrote, the most conclusive way of discrediting a political opponent is still to describe him as a disciple of Machiavelli.[2] Bacon was one of the first to praise him for 'saying openly and without hypocrisy what men are in the habit of doing, not what they ought to do'.[3] Henceforth no political thinker could ignore him. In France Bodin, in England Hobbes, in The Netherlands Spinoza, professed to find a compromise between the new doctrine and the conception of a 'law of nature' constituting a supreme ethical standard. But all three were in substance realists; and the age of Newton for the first time conceived the possibility of a physical science of politics.[4] The work of Bodin and Hobbes, writes Professor Laski, was 'to separate ethics from politics, and to complete by theoretical means the division which Machiavelli had effected on practical grounds'.[5] 'Before the names of Just and Unjust can have place', said Hobbes, 'there must be some coercive power.'[6] Spinoza believed that practical statesmen had contributed more to the understanding of politics than men of theory 'and, above all, theologians'; for 'they have put themselves to the school of experience, and have therefore taught nothing which does not bear upon our practical needs'.[7] In anticipation of Hegel, Spinoza declares that 'every man does what he does according to the laws of his nature and to the highest right of nature'.[8] The way is thus opened for determinism; and ethics become, in the last analysis, the study of reality.

Modern realism differs, however, in one important respect from that of the sixteenth and seventeenth centuries. Both utopianism and realism accepted and incorporated in their philosophies the eighteenth-century belief in progress, with the curious and somewhat paradoxical result that realism became in appearance more 'progressive' than utopianism. Utopianism grafted its belief in progress on to its belief in an absolute ethical standard, which remained *ex hypothesi* static. Realism, having no such sheet anchor, became more and more dynamic and relativist. Progress became part of the inner essence of the historical process; and mankind was moving forward towards a goal which was left undefined, or was differently defined by different philosophers. The 'historical school' of realists had its home in Germany, and its development is traced through the great names of Hegel and Marx. But no country in Western Europe, and no branch of thought, was immune from its influence in the middle and later years of the nineteenth century; and this development, while it has freed realism from the pessimistic colouring imparted to it by thinkers like Machiavelli and Hobbes, has thrown its determinist character into stronger relief.

The idea of causation in history is as old as the writing of history itself. But so long as the belief prevailed that human affairs were subject to the continuous supervision and occasional intervention of a Divine Providence, no philosophy of history based on a regular relationship of cause and effect was likely to be evolved. The substitution of reason for Divine Providence enabled Hegel to produce, for the first time, a philosophy based on the conception of a rational historical process. Hegel, while assuming a regular and orderly process, was content to find its directing force in a metaphysical abstraction – the *Zeitgeist*. But once the historical conception of reality had established itself, it was a short step to substitute for the abstract *Zeitgeist* some concrete material force. The economic interpretation of history was not invented, but developed and popularized, by Marx. About the same time Buckle propounded a geographical interpretation of history which convinced him that human affairs were 'permeated by one glorious principle of universal and undeviating regularity';[9] and this has been revived in the form of the science of *Geopolitik*, whose inventor describes geography as 'a political categorical imperative'.[10] Spengler believed that events were determined by quasi-biological laws governing the growth and decline of civilizations. More eclectic thinkers interpret history as the product of a variety of material factors, and the policy of a group or nation as a reflexion of all the material factors which make up the group or national interest. 'Foreign policies', said Mr Hughes during his tenure of office as American Secretary of State, 'are not built upon abstractions. They are the result of national interest arising from some immediate exigency or standing out vividly in historical perspective'.[11] Any such interpretation of reality, whether in terms of a *Zeitgeist*, or of economics or geography, or of 'historical perspective', is in its last analysis deterministic. Marx (though, having a programme of action, he could not be a rigid and consistent determinist) believed in 'tendencies which work out with an iron necessity towards an inevitable goal'.[12] 'Politics', wrote Lenin, 'have their own objective logic independent of the prescriptions of this or that individual or party.'[13] In January 1918, he described his belief in the coming socialist revolutions in Europe as 'a scientific prediction'.[14]

On the 'scientific' hypothesis of the realists, reality is thus identified with the whole course of historical evolution, whose laws it is the business of the philosopher to investigate and reveal. There can be no reality outside the historical process. 'To conceive of history as evolution and progress', writes Croce, 'implies accepting it as necessary in all its parts, and therefore denying validity to judgements on it.'[15] Condemnation of the past on ethical grounds has no meaning; for in Hegel's words, 'philosophy transfigures the real which appears unjust into the rational'.[16] What was, is right. History cannot be judged except by historical standards. It is significant that our historical judgements, except those relating to a past which we can ourselves remember as the present, always appear to start from the presupposition that things could not have turned out otherwise than they did. It is recorded that Venizelos, on reading in Fisher's *History of Europe* that the Greek invasion of Asia Minor in 1919 was a mistake, smiled ironically and said: 'Every enterprise that does not succeed is a mistake.'[17] If Wat Tyler's rebellion had succeeded, he would be an English national hero. If the American War of Independence had ended in disaster, the Founding Fathers of the United States would be briefly recorded in history as a gang of turbulent and unscrupulous fanatics. Nothing succeeds like success. 'World history', in the famous phrase which Hegel borrowed from Schiller, 'is the world court.' The popular paraphrase 'Might is Right' is misleading only if we attach too restricted a meaning to the word 'Might'. History creates rights, and therefore right. The doctrine of the survival of the fittest proves that the survivor was, in fact, the fittest to survive. Marx does not seem to have maintained that the victory of the proletariat was just in any other sense than that it was historically inevitable. Lukacs was a consistent, though perhaps indiscreet, Marxist when he based the 'right' of the proletariat on its 'historical mission'.[18] Hitler believed in the historical mission of the German people.

The relativity of thought

The outstanding achievement of modern realism, however, has been to reveal, not merely the determinist aspects of the historical process, but the relative and pragmatic character of thought itself. In the last fifty years, thanks mainly though not wholly to the influence of Marx, the principles of the historical school have been applied to the analysis of thought; and the foundations of a new science have been laid, principally by German thinkers, under the name of the 'sociology of knowledge'. The realist has thus been enabled to demonstrate that the intellectual theories and ethical standards of utopianism, far from being the expression of absolute and *a priori* principles, are historically conditioned, being both products of circumstances and interests and weapons framed for the furtherance of interests. 'Ethical notions', as Mr Bertrand Russell has remarked, 'are very seldom a cause, but almost always an effect, a means of claiming universal legislative authority for our own preference, not, as we fondly imagine, the actual ground of those preferences.'[19] This is by far the most formidable attack which utopianism has to face; for here the very foundations of its belief are undermined by the realist critique.

In a general way, the relativity of thought has long been recognized. As early as the seventeenth century Bishop Burnet expounded the relativist view as cogently, if not as pungently, as Marx:

> As to the late Civil Wars, 'tis pretty well known what notions of government went current in those days. When monarchy was to be subverted we knew what was necessary to justify the fact; and then, because it was convenient for the purpose, it was undoubtedly true in the nature of things that government had its original from the people, and the prince was only their trustee. ... But afterwards, when monarchy took its place again ... another notion of government came into fashion. Then government had its original entirely from God, and the prince was accountable to none but Him. ... And now, upon another turn of things, when people have a liberty to speak out, a new set of notions is advanced; now passive obedience is all a mistake, and instead of being a duty to suffer oppression, 'tis a glorious act to resist it: and instead of leaving injuries to be redressed by God, we have a natural right to relieve ourselves.[20]

In modern times, the recognition of this phenomenon has become fairly general. 'Belief, and to speak fairly, honest belief,' wrote Dicey of the divisions of opinion in the nineteenth century about slavery, 'was to a great extent the result not of argument, not even of direct self-interest, but of circumstances. ... Circumstances are the creators of most men's opinions.'[21] Marx narrowed down this somewhat vague conception, declaring that all thought was conditioned by the economic interest and social status of the thinker. This view was perhaps unduly restrictive. In particular Marx, who denied the existence of 'national' interests, under-estimated the potency of nationalism as a force conditioning the thought of the individual. But the peculiar concentration which he applied to the principle served to popularize it and drive it home. The relativity of thought to the interests and circumstances of the thinker has been far more extensively recognized and understood since Marx wrote.

The principle has an extremely wide field of application. It has become a commonplace to say that theories do not mould the course of events, but are invented to explain them. 'Empire precedes imperialism.'[22] Eighteenth-century England 'put into practice the policy of *laissez-faire* before it found a justification, or even an apparent justification, in the new doctrine';[23] and 'the virtual break-up of *laissez-faire* as a body of doctrine ... has followed, and not preceded, the decline of *laissez-faire* in the real world'.[24] The theory of 'socialism in a single country' promulgated in Soviet Russia in 1924 was manifestly a product of the failure of Soviet régimes to establish themselves in other countries.

But the development of abstract theory is often influenced by events which have no essential connexion with it at all.

In the story of political thought [writes a modern social thinker] events have been no less potent than arguments. The failure and success of institutions, the victories and defeats of countries identified with certain principles have repeatedly brought new strength and resolution to the adherents or opponents of these principles as the case might be in all lands. ... Philosophy as it exists on earth is the word of philosophers who, authority tells us, suffer as much from toothache as other mortals, and are, like others, open to the impression of near and striking events and to the seductions of intellectual fashion.[25]

Germany's dramatic rise to power in the sixties and seventies of the last century was impressive enough to make the leading British philosophers of the next generation – Caird, T. H. Green, Bosanquet, McTaggart – ardent Hegelians. Thereafter, the Kaiser's telegram to Kruger and the German naval programme spread the conviction among British thinkers that Hegel was a less good philosopher than had been supposed; and since 1914 no British philosopher of repute has ventured to sail under the Hegelian flag. After 1870, Stubbs and Freeman put early English history on a sound Teutonic basis, while even in France Fustel de Coulanges had an uphill struggle to defend the Latin origins of French civilization. During the past thirty years, English historians have been furtively engaged in making the Teutonic origins of England as inconspicuous as possible.

Nor is it only professional thinkers who are subject to such influences. Popular opinion is not less markedly dominated by them. The frivolity and immorality of French life was an established dogma in nineteenth-century Britain, which still remembered Napoleon. 'When I was young,' writes Mr Bertrand Russell, 'the French ate frogs and were called "froggies", but they apparently abandoned this practice when we concluded our *entente* with them in 1904 – at any rate, I have never heard it mentioned since that date.'[26] Some years later, 'the gallant little Jap' of 1905 underwent a converse metamorphosis into 'the Prussian of the East'. In the nineteenth century, it was a commonplace of British opinion that Germans were efficient and enlightened, and Russians backward and barbarous. About 1910, it was ascertained that Germans (who turned out to be mostly Prussians) were coarse, brutal and narrow-minded, and that Russians had a Slav soul. The vogue of Russian literature in Great Britain, which set in about the same time, was a direct outcome of the political *rapprochement* with Russia. The vogue of Marxism in Great Britain and France, which began on a modest scale after the success of the

Bolshevik revolution in Russia, rapidly gathered momentum, particularly among intellectuals, after 1934, when it was discovered that Soviet Russia was a potential military ally against Germany. It is symptomatic that most people, when challenged, will indignantly deny that they form their opinions in this way; for as Acton long ago observed, 'few discoveries are more irritating than those which expose the pedigree of ideas'.[27] The conditioning of thought is necessarily a subconscious process.

The adjustment of thought to purpose

Thought is not merely relative to the circumstances and interests of the thinker: it is also pragmatic in the sense that it is directed to the fulfilment of his purposes. For the realist, as a witty writer has put it, truth is 'no more than the perception of discordant experience pragmatically adjusted for a particular purpose and for the time being'.[28] The purposeful character of thought has been discussed in a previous chapter; and a few examples will suffice here to illustrate the importance of this phenomenon in international politics.

Theories designed to discredit an enemy or potential enemy are one of the commonest forms of purposeful thinking. To depict one's enemies or one's prospective victims as inferior beings in the sight of God has been a familiar technique at any rate since the days of the Old Testament. Racial theories, ancient and modern, belong to this category; for the rule of one people or class over another is always justified by a belief in the mental and moral inferiority of the ruled. In such theories, sexual abnormality and sexual offences are commonly imputed to the discredited race or group. Sexual depravity is imputed by the white American to the negro; by the white South African to the Kaffir; by the Anglo-Indian to the Hindu; and by the Nazi German to the Jew. The most popular and most absurd of the charges levelled against the Bolsheviks in the early days of the Russian revolution was that they advocated sexual promiscuity. Atrocity stories, among which offences of a sexual character predominate, are the familiar product of war. On the eve of their invasion of Abyssinia, the Italians issued an official Green Book of Abyssinian atrocities. 'The Italian Government,' as the Abyssinian delegate at Geneva correctly observed, 'having resolved to conquer and destroy Ethiopia, begins by giving Ethiopia a bad name.'[29]

But the phenomenon also appears in less crude forms which sometimes enable it to escape detection. The point was well made by Crowe in a Foreign Office minute of March 1908:

The German (formerly Prussian) Government has always been most remarkable for the pains it takes to create a feeling of intense and holy hatred against a country with which it contemplates the possibility of war. It is undoubtedly in this way that the frantic hatred of England as a monster of personified selfishness and greed and absolute want of conscience, which now animates Germany, has been nursed and fed.[30]

The diagnosis is accurate and penetrating. But it is strange that so acute a mind as Crowe's should not have perceived that he himself was at this time performing, for the limited audience of statesmen and officials to which he had access, precisely the same operation of which he accused the German Government; for a perusal of his memoranda and minutes of the period reveals an able, but transparent, attempt to 'create a feeling of intense and holy hatred' against his own country's future enemy – a curious instance of our promptness to detect the conditioned or purposeful character of other people's thought, while assuming that our own is wholly objective.

The converse of this propagation of theories designed to throw moral discredit on an enemy is the propagation of theories reflecting moral credit on oneself and one's own policies. Bismarck records the remark made to him by Walewski, the French Foreign Minister, in 1857, that it was the business of a diplomat to cloak the interests of his country in the language of universal justice. More recently, Mr Churchill told the House of Commons that 'there must be a moral basis for British rearmament and foreign policy'.[31] It is rare, however, for modern statesmen to express themselves with this frankness; and in contemporary British and American politics, the most powerful influence has been wielded by those more utopian statesmen who are sincerely convinced that policy is deduced from ethical principles, not ethical principles from policy. The realist is nevertheless obliged to uncover the hollowness of this conviction. 'The right', said Woodrow Wilson to the United States Congress in 1917, 'is more precious than peace.'[32] 'Peace comes before all,' said Briand ten years later to the League of Nations Assembly, 'peace comes even before justice.'[33] Considered as ethical principles, both these contradictory pronouncements are tenable and could muster respectable support. Are we therefore to believe that we are dealing with a clash of ethical standards, and that if Wilson's and Briand's policies differed it was because they deduced them from opposite principles? No serious student of politics will entertain this belief. The most cursory examination shows that the principles were deduced from the policies, not the policies from the principles. In 1917, Wilson had decided on the policy of war with Germany, and he proceeded to clothe that policy in the appropriate garment of righteousness. In 1928 Briand was fearful of attempts made in the name of justice to disturb a peace settlement favourable to France; and he had no more difficulty than Wilson in finding the moral phraseology which fitted his policy. It would be irrelevant to discuss this supposed difference of principles on ethical grounds. The principles merely reflected different national policies framed to meet different conditions.

The double process of morally discrediting the policy of a potential enemy and morally justifying one's own may be abundantly illustrated from the discussions of disarmament between the two wars. The experience of the Anglo-Saxon Powers, whose naval predominance had been threatened by the submarine, provided an ample opportunity of denouncing the immorality of this new weapon. 'Civilization demands', wrote the naval adviser to the American Delegation at the Peace Conference, 'that naval warfare be placed on a higher plane' by the abolition of the submarine.[34] Unfortunately the submarine was regarded as a convenient weapon by the weaker French, Italian and Japanese navies; and this particular demand of civilization could not therefore be complied with. A distinction of a more sweeping character was established by Lord Cecil in a speech to the General Council of the League of Nations Union in 1922:

> The general peace of the world will not be materially secured merely by naval disarmament. .... If all the maritime Powers were to disarm, or drastically limit their armaments, I am not at all sure that would not increase the danger of war rather than decrease it, because the naval arm is mainly defensive; the offensive must be to a large extent the military weapon.[35]

The inspiration of regarding one's own vital armaments as defensive and beneficent and those of other nations as offensive and wicked proved particularly fruitful. Exactly ten years later, three commissions of the Disarmament Conference spent many weeks in a vain endeavour to classify armaments as 'offensive' and 'defensive'. Delegates of all nations showed extraordinary ingenuity in devising arguments, supposedly based on pure objective theory, to prove that the armaments on which they chiefly relied were defensive, while those of potential rivals were essentially offensive. Similar attitudes have been taken up in regard to economic 'armaments'. In the latter part of the nineteenth century – and in a lesser degree down to 1931 – protective tariffs were commonly regarded in Great Britain as immoral. After 1931 straight tariffs regained their innocence, but barter agreements, industrial (though not agricultural) quotas, exchange controls and other weapons employed by Continental states were still tainted with immorality. Down to 1930, successive revisions of the United States tariff had almost invariably been upward; and American economists, in other respects staunch upholders of *laissez-faire*, had almost invariably treated tariffs as legitimate and laudable. But the change in the position of the United States from a debtor to a creditor Power, combined with the reversal of British economic policy, altered the picture; and the reduction of tariff barriers has come to be commonly identified by American spokesmen with the cause of international morality.

National interest and the universal good

The realist should not, however, linger over the infliction of these pinpricks through chinks in the utopian defences. His task is to bring down the whole cardboard structure of utopian thought by exposing the hollowness of the material out of which it is built. The weapon of the relativity of thought must be used to demolish the utopian concept of a fixed and absolute standard by which policies and actions can be judged. If theories are revealed as a reflexion of practice and principles of political needs, this discovery will apply to the fundamental theories and principles of the utopian creed, and not least to the doctrine of the harmony of interests which is its essential postulate.

It will not be difficult to show that the utopian, when he preaches the doctrine of the harmony of interests, is innocently and unconsciously adopting Walewski's maxim, and clothing his own interest in the guise of a universal interest for the purpose of imposing it on the rest of the world. 'Men come easily to believe that arrangements agreeable to themselves are beneficial to others,' as Dicey observed;[36] and theories of the public good, which turn out on inspection to be an elegant disguise for some particular interest, are as common in international as in national affairs. The utopian, however eager he may be to establish an absolute standard, does not argue that it is the duty of his country, in conformity with that standard, to put the interest of the world at large before its own interest; for that would be contrary to his theory that the interest of all coincides with the interest of each. He argues that what is best for the world is best for his country, and then reverses the argument to read that what is best for his country is best for the world, the two propositions being, from the utopian standpoint, identical; and this unconscious cynicism of the contemporary utopian has proved a far more effective diplomatic weapon than the deliberate and self-conscious cynicism of a Walewski or a Bismarck. British writers of the past half-century have been particularly eloquent supporters of the theory that the maintenance of British supremacy is the performance of a duty to mankind. 'If Great Britain has turned itself into a coal-shed and blacksmith's forge,' remarked *The Times* ingenuously in 1885, 'it is for the behoof of mankind as well as its own.'³⁷ The following extract is typical of a dozen which might be culled from memoirs of public men of the period:

> I have but one great object in this world, and that is to maintain the greatness of the Empire. But apart from my John Bull sentiment upon the point, I firmly believe that in doing so I work in the cause of Christianity, of peace, of civilisation, and the happiness of the human race generally.³⁸

'I contend that we are the first race in the world,' wrote Cecil Rhodes, 'and that the more of the world we inhabit the better it is for the human race.'³⁹ In 1891, the most popular and brilliant journalist of the day, W. T. Stead, founded the *Review of Reviews*. 'We believe in God, in England and in Humanity,' ran the editorial manifesto in its opening number. 'The English-speaking race is one of the chief of God's chosen agents for executing coming improvements in the lot of mankind.'⁴⁰ An Oxford professor was convinced in 1912 that the secret of Britain's history was that 'in fighting for her own independence she has been fighting for the freedom of Europe, and that the service thus rendered to Europe and to mankind has carried with it the possibility of that larger service to which we give the name Empire'.⁴¹

The first world war carried this conviction to a pitch of emotional frenzy. A bare catalogue, culled from the speeches of British statesmen, of the services which British belligerency was rendering to humanity would fill many pages. In 1917, Balfour told the New York Chamber of Commerce that 'since August 1914, the fight has been for the highest spiritual advantages of mankind, without a petty thought or ambition'.⁴² The Peace Conference and its sequel temporarily discredited these professions and threw some passing doubt on the belief in British supremacy as one of the moral assets of mankind. But the period of disillusionment and modesty was short. Moments of international tension, and especially moments when the possibility of war appears on the horizon, always stimulate this identification of national interest with morality. At the height of the Abyssinian crisis, the Archbishop of Canterbury admonished the French public through an interview in a Paris newspaper:

> We are animated by moral and spiritual considerations. I do not think I am departing from my role by contributing towards the clearing up of this misunderstanding...
>   It is ... no egoist interest that is driving us forward, and no consideration of interest should keep you behind.⁴³

In the following year, Professor Toynbee was once more able to discover that the security of the British Empire 'was also the supreme interest of the whole world'.[44] In 1937, Lord Cecil spoke to the General Council of the League of Nations Union of 'our duty to our country, to our Empire and to humanity at large', and quoted:

> Not once nor twice in our rough island story
> The path of duty is the way to glory.[45]

An Englishman, as Mr Bernard Shaw remarks in *The Man of Destiny*, 'never forgets that the nation which lets its duty get on to the opposite side to its interest is lost'. It is not surprising that an American critic should recently have described the British as 'Jesuits lost to the theological but gained for the political realm',[46] or that a former Italian Minister for Foreign Affairs should have commented, long before these recent manifestations, on 'that precious gift bestowed upon the British people – the possession of writers and clergymen able in perfect good faith to advance the highest moral reasons for the most concrete diplomatic action, with inevitable moral profit to England'.[47]

In recent times, the same phenomenon has become endemic in the United States. The story of how McKinley prayed for divine guidance and decided to annex the Philippines is a classic of modern American history; and this annexation was the occasion of a popular outburst of moral self-approval hitherto more familiar in the foreign policy of Great Britain than of the United States. Theodore Roosevelt, who believed more firmly than any previous American President in the doctrine *L'état, c'est moi*, carried the process a step further. The following curious dialogue occurred in his cross-examination during a libel action brought against him in 1915 by a Tammany leader:

Query: How did you know that substantial justice was done?
ROOSEVELT: Because I did it, because ... I was doing my best.
Query: You mean to say that, when you do a thing, thereby substantial justice is done.
ROOSEVELT: I do. When I do a thing, I do it so as to do substantial justice. I mean just that.[48]

Woodrow Wilson was less naïvely egotistical, but more profoundly confident of the identity of American policy and universal justice. After the bombardment of Vera Cruz in 1914, he assured the world that 'the United States had gone down to Mexico to serve mankind'.[49] During the first world war, he advised American naval cadets 'not only always to think first of America, but always, also, to think first of humanity' – a feat rendered slightly less difficult by his explanation that the United States had been 'founded for the benefit of humanity'.[50] Shortly before the entry of the United States into the war, in an address to the Senate on war aims, he stated the identification still more categorically: 'These are American principles, American policies ... They are the principles of mankind and must prevail.'[51]

It will be observed that utterances of this character proceed almost exclusively from Anglo-Saxon statesmen and writers. It is true that when a prominent National Socialist asserted that 'anything that benefits the German people is right, anything that harms the German people is wrong',[52] he was merely propounding the same identification of national interest with universal right which had already been established for English-speaking countries by Wilson, Professor Toynbee, Lord Cecil and many others. But when the claim is translated into a foreign language, the note seems forced, and the identification unconvincing, even to the peoples concerned. Two explanations are commonly given of this curious discrepancy. The first explanation, which is popular in English-speaking countries, is that the policies of the English-speaking nations are in fact more virtuous and disinterested than those of Continental states, so that Wilson and Professor Toynbee and Lord Cecil are, broadly speaking, right when they identify the American and British national interest with the interest of mankind. The second explanation, which is popular in Continental countries, is that the English-speaking peoples are past masters in the art of concealing their selfish national interests in the guise of the general good, and that this kind of hypocrisy is a special and characteristic peculiarity of the Anglo-Saxon mind.

It seems unnecessary to accept either of these heroic attempts to cut the knot. The solution is a simple one. Theories of social morality are always the product of a dominant group which identifies itself with the community as a whole, and which possesses facilities denied to subordinate groups or individuals for imposing its view of life on the community. Theories of international morality are, for the same reason and in virtue of the same process, the product of dominant nations or groups of nations. For the past hundred years, and more especially since 1918, the English-speaking peoples have formed the dominant group in the world; and current theories of international morality have been designed to perpetuate their supremacy and expressed in the idiom peculiar to them. France, retaining something of her eighteenth-century tradition and restored to a position of dominance for a short period after 1918, has played a minor part in the creation of current international morality, mainly through her insistence on the role of law in the moral order. Germany, never a dominant Power and reduced to helplessness after 1918, has remained for these reasons outside the charmed circle of creators of international morality. Both the view that the English-speaking peoples are monopolists of international morality and the view that they are consummate international hypocrites may be reduced to the plain fact that the current canons of international virtue have, by a natural and inevitable process, been mainly created by them.

The realist critique of the harmony of interests

The doctrine of the harmony of interests yields readily to analysis in terms of this principle. It is the natural assumption of a prosperous and privileged class, whose members have a dominant voice in the community and are therefore naturally prone to identify its interests with their own. In virtue of this identification, any assailant of the interests of the dominant group is made to incur the odium of assailing the alleged common interest of the whole community, and is told that in making this assault he is attacking his own higher interests. The doctrine of the harmony of interests thus serves as an ingenious moral device invoked, in perfect sincerity, by privileged groups in order to justify and maintain their dominant position. But a further point requires notice. The supremacy within the community of the privileged group may be, and often is, so overwhelming that there is, in fact, a sense in which its interests are those of the community, since its well-being necessarily carries with it some measure of well-being for other members of the community, and its collapse would entail the collapse of the community as a whole. In so far, therefore, as the alleged natural harmony of interests has any reality, it is created by the overwhelming power of the privileged group, and is an excellent illustration of the Machiavellian maxim that morality is the product of power. A few examples will make this analysis of the doctrine of the harmony of interests clear.

In the nineteenth century, the British manufacturer or merchant, having discovered that *laissez-faire* promoted his own prosperity, was sincerely convinced that it also promoted British prosperity as a whole. Nor was this alleged harmony of interests between himself and the community entirely fictitious. The predominance of the manufacturer and the merchant was so overwhelming that there was a sense in which an identity between their prosperity and British prosperity as a whole could be correctly asserted. From this it was only a short step to argue that a worker on strike, in damaging the prosperity of the British manufacturer, was damaging British prosperity as a whole, and thereby damaging his own, so that he could be plausibly denounced by the predecessors of Professor Toynbee as immoral and by the predecessors of Professor Zimmern as muddle-headed. Moreover, there was a sense in which this argument was perfectly correct. Nevertheless, the doctrine of the harmony of interests and of solidarity between the classes must have seemed a bitter mockery to the underprivileged worker, whose inferior status and insignificant stake in 'British prosperity' were consecrated by it; and presently he was strong enough to force the abandonment of *laissez-faire* and the substitution for it of the 'social service state', which implicitly denies the natural harmony of interests and sets out to create a new harmony by artificial means.

The same analysis may be applied in international relations. British nineteenth-century statesmen, having discovered that free trade promoted British prosperity, were sincerely convinced that, in doing so, it also promoted the prosperity of the world as a whole. British predominance in world trade was at that time so overwhelming that there was a certain undeniable harmony between British interests and the interests of the world. British prosperity flowed over into other countries, and a British economic collapse would have meant worldwide ruin. British free traders could and did argue that protectionist countries were not only egotistically damaging the prosperity of the world as a whole, but were stupidly damaging their own, so that their behaviour was both immoral and muddle-headed. In British eyes, it was irrefutably proved that international trade was a single whole, and flourished or slumped together. Nevertheless, this alleged international harmony of interests seemed a mockery to those underprivileged nations whose inferior status and insignificant stake in international trade were consecrated by it. The revolt against it destroyed that overwhelming British preponderance which had provided a plausible basis for the theory. Economically, Great Britain in the nineteenth century was dominant enough to make a bold bid to impose on the world her own conception of international economic morality. When competition of all against all replaced the domination of the world market by a single Power, conceptions of international economic morality necessarily became chaotic.

Politically, the alleged community of interest in the maintenance of peace, whose ambiguous character has already been discussed, is capitalized in the same way by a dominant nation or group of nations. Just as the ruling class in a community prays for domestic peace, which guarantees its own security and predominance, and denounces class war, which might threaten them, so international peace becomes a special vested interest of predominant Powers. In the past, Roman and British imperialism were commended to the world in the guise of the *pax Romana* and the *pax Britannica*. To-day, when no single Power is strong enough to dominate the world, and supremacy is vested in a group of nations, slogans like 'collective security' and 'resistance to aggression' serve the same purpose of proclaiming an identity of interest between the dominant group and the world as a whole in the maintenance of peace. Moreover, as in the examples we have just considered, so long as the supremacy of the dominant group is sufficiently great, there is a sense in which this identity of interest exists. 'England', wrote a German professor in the nineteen-twenties, 'is a solitary

Power with a national programme which, while egotistic through and through, at the same time promises to the world something which the world passionately desires: order, progress and eternal peace.'[53] When Mr Churchill declared that 'the fortunes of the British Empire and its glory are inseparably interwoven with the fortunes of the world',[54] this statement had precisely the same foundation in fact as the statement that the prosperity of British manufacturers in the nineteenth century was inseparably interwoven with British prosperity as a whole. Moreover, the purpose of the statements was precisely the same, namely to establish the principle that the defence of the British Empire, or the prosperity of the British manufacturer, was a matter of common interest to the whole community, and that anyone who attacked it was therefore either immoral or muddle-headed. It is a familiar tactic of the privileged to throw moral discredit on the underprivileged by depicting them as disturbers of the peace; and this tactic is as readily applied internationally as within the national community. 'International law and order', writes Professor Toynbee of a recent crisis, 'were in the true interests of the whole of mankind ... whereas the desire to perpetuate the region of violence in international affairs was an anti-social desire which was not even in the ultimate interests of the citizens of the handful of states that officially professed this benighted and anachronistic creed.'[55] This is precisely the argument, compounded of platitude and falsehood in about equal parts, which did duty in every strike in the early days of the British and American Labour movements. It was common form for employers, supported by the whole capitalist press, to denounce the 'anti-social' attitude of trade union leaders, to accuse them of attacking law and order and of introducing 'the reign of violence', and to declare that 'true' and 'ultimate' interests of the workers lay in peaceful co-operation with the employers.[56] In the field of social relations, the disingenuous character of this argument has long been recognized. But just as the threat of class war by the proletarian is 'a natural cynical reaction to the sentimental and dishonest efforts of the privileged classes to obscure the conflict of interest between classes by a constant emphasis on the minimum interests which they have in common',[57] so the warmongering of the dissatisfied Powers was the 'natural, cynical reaction' to the sentimental and dishonest platitudinizing of the satisfied Powers on the common interest in peace. When Hitler refused to believe 'that God has permitted some nations first to acquire a world by force and then to defend this robbery with moralizing theories',[58] he was merely echoing in another context the Marxist denial of a community of interest between 'haves' and 'have-nots', the Marxist exposure of the interested character of '*bourgeois* morality', and the Marxist demand for the expropriation of the expropriators.

The crisis of September 1938 demonstrated in a striking way the political implications of the assertion of a common interest in peace. When Briand proclaimed that 'peace comes before all', or Mr Eden that 'there is no dispute which cannot be settled by peaceful means',[59] the assumption underlying these platitudes was that, so long as peace was maintained, no changes distasteful to France or Great Britain could be made in the *status quo*. In 1938, France and Great Britain were trapped by the slogans which they themselves had used in the past to discredit the dissatisfied Powers, and Germany had become sufficiently dominant (as France and Great Britain had hitherto been) to turn the desire for peace to her own advantage. About this time, a significant change occurred in the attitude of the German and Italian dictators. Hitler eagerly depicted Germany as a bulwark of peace menaced by warmongering democracies. The League of Nations, he declared in his Reichstag speech of 28 April 1939, is a 'stirrer up of trouble', and collective security means 'continuous danger of war'. Mussolini borrowed the British formula about the possibility of settling all international disputes by peaceful means, and declared that 'there are not in Europe at present problems so big and so active as to justify a war which from a European conflict would naturally become universal'.[60] Such utterances were symptoms that Germany and Italy were already looking forward to the time when, as dominant Powers, they would acquire the vested interest in peace recently enjoyed by Great Britain and France, and be able to get their way by pillorying the democratic countries as enemies of peace. These developments may have made it easier to appreciate Halévy's subtle observation that 'propaganda against war is itself a form of war propaganda'.[61]

The realist critique of internationalism

The concept of internationalism is a special form of the doctrine of the harmony of interests. It yields to the same analysis; and there are the same difficulties about regarding it as an absolute standard independent of the interests and policies of those who promulgate it. 'Cosmopolitanism', wrote Sun Yat-sen, 'is the same thing as China's theory of world empire two thousand years ago ... China once wanted to be sovereign lord of the earth and to stand above every other nation, so she espoused cosmopolitanism.'[62] In the Egypt of the Eighteenth Dynasty, according to Freud, 'imperialism was reflected in religion as universality and monotheism'.[63] The doctrine of a single world-state, propagated by the Roman Empire and later by the Catholic Church, was the symbol of a claim to universal dominion. Modern internationalism has its genesis in seventeenth- and eighteenth-century France, during which French hegemony in Europe was at its height. This was the period which produced Sully's *Grand Dessin* and the Abbé Saint-Pierre's *Projet de Paix Perpétuelle* (both plans to perpetuate an international *status quo* favourable to the French monarchy), which saw the birth of the humanitarian and cosmopolitan doctrines of the Enlightenment, and which established French as the universal language of educated people. In the next century, the leadership passed to Great Britain, which became the home of internationalism. On the eve of the Great Exhibition of 1851 which, more than any other single event, established Great Britain's title to world supremacy, the Prince Consort spoke movingly of 'that great end to which ... all history points – the realisation of the unity of mankind';[64] and Tennyson hymned 'the parliament of man, the federation of the world'. France chose the moment of her greatest supremacy in the nineteen-twenties to launch a plan of 'European Union'; and Japan shortly afterwards developed an ambition to proclaim herself the leader of a united Asia. It was symptomatic of the growing international predominance of the United States when widespread popularity was enjoyed in the late nineteen-thirties by the book of an American journalist advocating a world union of democracies, in which the United States would play the predominant role.[65]

Just as pleas for 'national solidarity' in domestic politics always come from a dominant group which can use this solidarity to strengthen its own control over the nation as a whole, so pleas for international solidarity and world union come from those dominant nations which may hope to exercise control over a unified world. Countries which are struggling to force their way into the dominant group naturally tend to invoke nationalism against the internationalism of the controlling Powers. In the sixteenth century, England opposed her nascent nationalism to the internationalism of the Papacy and the Empire. In the past century and a half Germany opposed her nascent nationalism to the internationalism first of France, then of Great Britain. This circumstance made her impervious to those universalist and humanitarian doctrines which were popular in eighteenth-century France and nineteenth-century Britain; and her hostility to internationalism was further aggravated after 1919, when Great Britain and France endeavoured to create a new 'international order' as a bulwark of their own predominance. 'By "international",' wrote a German correspondent in *The Times*, 'we have come to understand a conception that places other nations at an advantage over our own.'[66] Nevertheless, there was little doubt that Germany, if she became supreme in Europe, would adopt international slogans and establish some kind of international organization to bolster up her power. A British Labour ex-Minister at one moment advocated the suppression of Article 16 of the Covenant of the League of Nations on the unexpected ground that the totalitarian states might some day capture the League and invoke that article to justify the use of force by themselves.[67] It seemed more likely that they would seek to develop the Anti-Comintern Pact into some form of international organization. 'The Anti-Comintern Pact', said Hitler in the Reichstag on 30 January 1939, 'will perhaps one day become the crystallization point of a group of Powers whose ultimate aim is none other than to eliminate the menace to the peace and culture of the world instigated by a satanic apparition.' 'Either Europe must achieve solidarity,' remarked an Italian journal about the same time, 'or the "axis" will impose it.'[68] 'Europe in its entirety', said Goebbels, 'is adopting a new order and a new orientation under the intellectual leadership of National Socialist Germany and Fascist Italy.'[69] These were symptoms not of a change of heart, but of the fact that Germany and Italy felt themselves to be approaching the time when they might become strong enough to espouse internationalism. 'International order' and 'international solidarity' will always be slogans of those who feel strong enough to impose them on others.

The exposure of the real basis of the professedly abstract principles commonly invoked in international politics is the most damning and most convincing part of the realist indictment of utopianism. The nature of the charge is frequently misunderstood by those who seek to refute it. The charge is not that human beings fail to live up to their principles. It matters little that Wilson, who thought that the right was more precious than peace, and Briand, who thought that peace came even before justice, and Mr Eden, who believed in collective security, failed themselves, or failed to induce their countrymen, to apply these principles consistently. What matters is that these supposedly absolute and universal principles were not principles at all, but the unconscious reflexions of national policy based on a particular interpretation of national interest at a particular time. There is a sense in which peace and co-operation between nations or classes or individuals is a common and universal end irrespective of conflicting interests and politics. There is a sense in which a common interest exists in the maintenance of order, whether it be international order or 'law and order' within the nation. But as soon as the attempt is made to apply these supposedly abstract principles to a concrete political situation, they are revealed as the transparent disguises of selfish vested interests. The bankruptcy of utopianism resides not in its failure to live up to its principles, but in the exposure of its inability to provide any absolute and disinterested standard for the conduct of international affairs. The utopian, faced by the collapse of standards whose interested character he has failed to penetrate, takes refuge in condemnation of a reality which refuses to conform to these standards. A passage penned by the German historian

Meinecke after the first world war is the best judgement by anticipation of the role of utopianism in the international politics of the period:

> The profound defect of the Western, natural-law type of thought was that, when applied to the real life of the state, it remained a dead letter, did not penetrate the consciousness of statesmen, did not hinder the modern hypertrophy of state interest, and so led either to aimless complaints and doctrinaire suppositions or else to inner falsehood and cant.[70]

These 'aimless complaints', these 'doctrinaire suppositions', this 'inner falsehood and cant' will be familiar to all those who have studied what was written about international politics in English-speaking countries between the two world wars.

Notes

1. Machiavelli, *The Prince*, chs. 15 and 23 (Engl. transl., Everyman's Library, pp. 121, 193).
2. Two curious recent illustrations may be cited. In the chapter of the *Survey of International Affairs* dealing with the Nazi revolution, Professor Toynbee declares that National Socialism is the 'fulfilment of ideals ... formulated ... by Machiavelli'; and he reiterates this view in two further passages of considerable length in the same chapter (*Survey of International Affairs, 1934*, pp. 111, 117–19, 126–8). In the trial of Zinoviev, Kamenev and others in Moscow in August 1936, the Public Prosecutor, Vyshinsky, quoted a passage from Kamenev's writings in which Machiavelli had been praised as 'a master of political aphorism and a brilliant dialectician', and accused Kamenev of having 'adopted the rules of Machiavelli' and 'developed them to the utmost point of unscrupulousness and immorality' (*The Case of the Trotskyite-Zinovievite Centre*, pp. 138–9).
3. Bacon, *On the Advancement of Learning*, vii. ch. 2.
4. Hobbes's scheme, 'there was in theory no place for any new force or principle beyond the laws of motion found at the beginning; there were merely complex cases of mechanical causation' (Sabine, *History of Political Thought*, p. 458)
5. Introduction to *A Defence of Liberty against Tyrants* (*Vindiciae contra Tyrannos*), ed. Laski, p. 45.
6. Hobbes, *Leviathan*, ch. xv.
7. Spinoza, *Tractatus Politicus*, i. pp. 2–3.
8. *Ibid.*, Introduction.
9. The concluding words of Buckle's *History of Civilisation*.
10. Kjellen, *Der Staat als Lebensform*, p. 81. Compare the opening words of Crowe's famous memorandum on British foreign policy: 'The general character of England's foreign policy is determined by the immutable conditions of her geographical situation' (*British Documents on the Origin of the War*, ed. Gooch and Temperley, iii. p. 397).

11. *International Conciliation*, No. 194, January 1924, p. 3.
12. Marx, *Capital*, Preface to 1st ed. (Engl. transl., Everyman's Library, p. 863).
13. Lenin, *Works* (2nd Russian ed.) x. p. 207.
14. *Ibid.*, xxii, p. 194.
15. Croce, *Storia della storiografia italiana*, i. p. 26.
16. Hegel, *Philosophie der Weltgeschichte* (Lasson's ed.), p. 55.
17. *Conciliation Internationale*, No. 5–6, 1937, p. 520.
18. Lukacs, *Geschichte und Klassenhewusstsein*, p. 215.
19. *Proceedings of the Aristotelian Society*, 1915–16, p. 302.
20. Burnet, *Essay upon Government*, p. 10.
21. Dicey, *Law and Opinion* (1905 ed.), p. 27.
22. J. A. Hobson, *Free Thought in the Social Sciences*, p. 190.
23. Halévy, *The Growth of Philosophic Radicalism* (Engl. transl.), p. 104.
24. M. Dobb, *Political Economy and Capitalism*, p. 188.
25. L. T. Hobhouse, *The Unity of Western Civilisation*, ed. F. S. Marvin (3rd ed.), pp. 177–8.
26. Bertrand Russell, *Which Way Peace?*, p. 158.
27. Acton, *History of Freedom*, p. 62.
28. Carl Becker, *Yale Review*, xxvii, p. 461.
29. *League of Nations: Official Journal*, November 1935, p. 1140.
30. *British Documents on the Origins of the War*, ed. Gooch and Temperley, vi. p. 131.
31. House of Commons, 14 March 1938: *Official Report*, cols. 95–9.
32. *The Public Papers of Woodrow Wilson: War and Peace*, ed. R. S. Baker, i. p. 16.
33. *League of Nations: Ninth Assembly*, p. 83.
34. R. S. Baker, *Woodrow Wilson and World Settlement*, iii, p. 120. There is an amusing nineteenth-century parallel. 'Privateering', wrote Queen Victoria at the time of the Conference of Paris in 1856, 'is a kind of Piracy which disgraces our civilisation; its abolition throughout the whole world would be a great step in advance.' We are not surprised to read that 'the privateer was then, like the submarine in modern times, the weapon of the weaker naval Power' (Sir William Malkin, *British Year Book of International Law*, viii. pp. 6, 30).
35. Published as League of Nations Union Pamphlet No. 76, p. 8. The very word 'militarism' conveys to most English readers the same connotation of the peculiar wickedness of armies. It was left to an American historian, Dr W. L. Langer, to coin the counterpart 'navalism', which has won significantly little acceptance.
36. Dicey, *Law and Opinion in England* (2nd ed.), pp. 14–15.
37. *The Times*, 27 August 1885.
38. Maurice and Arthur, *The Life of Lord Wolseley*, p. 314.
39. W. T. Stead, *The Last Will and Testament of Cecil J. Rhodes*, p. 58.
40. *Review of Reviews*, 15 January 1891.
41. Spencer Wilkinson, *Government and the War*, p. 116.
42. Quoted in Beard, *The Rise of American Civilisation*, ii. p. 646.
43. Quoted in *Manchester Guardian*, 18 October 1935.
44. Toynbee, *Survey of International Affairs*, 1935, ii. p. 46.
45. *Headway*, November 1937.
46. Carl Becker, *Yale Review*, xxvii. p. 452.
47. Count Sforza, *Foreign Affairs*, October 1927, p. 67.
48. Quoted in H. F. Pringle, *Theodore Roosevelt*, p. 318.

49. *Public Papers of Woodrow Wilson: The New Democracy*, ed. R. S. Baker, i. p. 104.
50. *Public Papers of Woodrow Wilson: The New Democracy*, ed. R. S. Baker, i. pp. 318–19.
51. *Ibid.*, ii. p. 414.
52. Quoted in Toynbee, *Survey of International Affairs*, 1936, p. 319.
53. Dibelius, *England*, p. 109.
54. Winston Churchill, *Arms and the Covenant*, p. 272.
55. Toynbee, *Survey of International Affair*, 1935, ii. p. 46.
56. 'Pray earnestly that right may triumph', said the representative of the Philadelphia coal-owners in an early strike organised by the United Mine Workers, 'remembering that the Lord God Omnipotent still reigns, and that His reign is one of law and order, and not of violence and crime' (H. F. Pringle, *Theodore Roosevelt*, p. 267).
57. R. Niebuhr, *Moral Man and Immoral Society*, p. 153.
58. Speech in the Reichstag, 30 January 1939.
59. *League of Nations: Eighteenth Assembly*, p. 63.
60. *The Times*, 15 May 1939.
61. Halévy, *A History of the English People in 1895–1905* (Engl. transl.), i. Introduction, p. xi.
62. Sun Yat-sen, *San Min Chu I* (Engl. transl.), pp. 68–9.
63. Sigmund Freud, *Moses and Monotheism*, p. 36.
64. T. Martin, *Life of the Prince Consort*, iii. p. 247.
65. Clarence Streit, *Union Now*.
66. *The Times*, 5 November 1938.
67. Lord Marley in the House of Lords, 30 November 1938: *Official Report*, col. 258.
68. *Relazioni Internazionali*, quoted in *The Times*, 5 December 1938.
69. *Völkischer Beobachter*, 1 April 1939.
70. Meinecke, *Staatsräson*, p. 533.

CHAPTER SIX

The Limitations of Realism

The exposure by realist criticism of the hollowness of the utopian edifice is the first task of the political thinker. It is only when the sham has been demolished that there can be any hope of raising a more solid structure in its place. But we cannot ultimately find a resting place in pure realism; for realism, though logically overwhelming, does not provide us with the springs of action which are necessary even to the pursuit of thought. Indeed, realism itself, if we attack it with its own weapons, often turns out in practice to be just as much conditioned as any other mode of thought. In politics, the belief that certain facts are unalterable or certain trends irresistible commonly reflects a lack of desire or lack of interest to change or resist them. The impossibility of being a consistent and thorough-going realist is one of the most certain and most curious lessons of political science. Consistent realism excludes four things which appear to be essential ingredients of all effective political thinking: a finite goal, an emotional appeal, a right of moral judgement and a ground for action.

The conception of politics as an infinite process seems in the long run uncongenial or incomprehensible to the human mind. Every political thinker who wishes to make an appeal to his contemporaries is consciously or unconsciously led to posit a finite goal. Treitschke declared that the 'terrible thing' about Machiavelli's teaching was 'not the immorality of the methods he recommends, but the lack of content of the state, which exists only in order to exist'.[1] In fact, Machiavelli is not so consistent. His realism breaks down in the last chapter of *The Prince*, which is entitled 'An Exhortation to free Italy from the Barbarians' – a goal whose necessity could be deduced from no realist premise. Marx, having dissolved human thought and action into the relativism of the dialectic, postulates the absolute goal of a classless society where the dialectic no longer operates – that one far-off event towards which, in true Victorian fashion, he believed the whole creation to be moving. The realist thus ends by negating his own postulate and assuming an ultimate reality outside the historical process. Engels was one of the first to level this charge against Hegel. 'The whole dogmatic content of the Hegelian system is declared to be absolute truth in contradiction to his dialectical method, which dissolves all dogmatism.'[2]

But Marx lays himself open to precisely the same criticism when he brings the process of dialectical materialism to an end with the victory of the proletariat. Thus utopianism penetrates the citadel of realism; and to envisage a continuing, but not infinite, process towards a finite goal is shown to be a condition of political thought. The greater the emotional stress, the nearer and more concrete is the goal. The first world war was rendered tolerable by the belief that it was the last of wars. Woodrow Wilson's moral authority was built up on the conviction, shared by himself, that he possessed the key to a just, comprehensive and final settlement of the political ills of mankind. It is noteworthy that almost all religions agree in postulating an ultimate state of complete blessedness.

The finite goal, assuming the character of an apocalyptic vision, thereby acquires an emotional, irrational appeal which realism itself cannot justify or explain. Everyone knows Marx's famous prediction of the future classless paradise:

> When work ceases to be merely a means of life and becomes the first living need; when, with the all-round development of the individual, productive forces also develop, and all the sources of collective wealth flow in free abundance – then only will it be possible to transcend completely the narrow horizon of *bourgeois* right, and society can inscribe on its banner: From each according to his capacities, to each according to his needs.[3]

Sorel proclaimed the necessity of a 'myth' to make revolutionary teaching effective; and Soviet Russia has exploited for this purpose the myth, first of world revolution, and more recently of the 'socialist fatherland'. There is much to be said for Professor Laski's view that 'communism has made its way by its idealism, and not by its realism, by its spiritual promise, not by its materialistic prospects'.[4] A modern theologian has analysed the situation with almost cynical clear-sightedness:

> Without the ultrarational hopes and passions of religion, no society will have the courage to conquer despair and attempt the impossible; for the vision of a just society is an impossible one, which can be approximated only by those who do not regard it as impossible. The truest visions of religion are illusions, which may be partly realized by being resolutely believed.[5]

And this again closely echoes a passage in *Mein Kampf* in which Hitler contrasts the 'programme-maker' with the politician:

> His [i.e. the programme-maker's] significance lies almost wholly in the future, and he is often what one means by the word '*weltfremd*'

[unpractical, utopian]. For if the art of the politician is really the art of the possible, then the programme-maker belongs to those of whom it is said that they please the gods only if they ask and demand from them the impossible.[6]

Credo quia impossible becomes a category of political thinking.

Consistent realism, as has already been noted, involves acceptance of the whole historical process and precludes moral judgements on it. As we have seen, men are generally prepared to accept the judgement of history on the past, praising success and condemning failure. This test is also widely applied to contemporary politics. Such institutions as the League of Nations, or the Soviet or Fascist régimes, are to a considerable extent judged by their capacity to achieve what they profess to achieve; and the legitimacy of this test is implicitly admitted by their own propaganda, which constantly seeks to exaggerate their successes and minimize their failures. Yet it is clear that mankind as a whole is not prepared to accept this rational test as a universally valid basis of political judgement. The belief that whatever succeeds is right, and has only to be understood to be approved, must, if consistently held, empty thought of purpose, and thereby sterilize and ultimately destroy it. Nor do those whose philosophy appears to exclude the possibility of moral judgements in fact refrain from pronouncing them. Frederick the Great, having explained that treaties should be observed for the reason that 'one can trick only once', goes on to call the breaking of treaties 'a bad and knavish policy', though there is nothing in his thesis to justify the moral epithet.[7] Marx, whose philosophy appeared to demonstrate that capitalists could only act in a certain way, spends many pages – some of the most effective in *Capital* – in denouncing the wickedness of capitalists for behaving in precisely that way. The necessity, recognized by all politicians, both in domestic and in international affairs, for cloaking interests in a guise of moral principles is in itself a symptom of the inadequacy of realism. Every age claims the right to create its own values, and to pass judgements in the light of them; and even if it uses realist weapons to dissolve other values, it still believes in the absolute character of its own. It refuses to accept the implication of realism that the word 'ought' is meaningless.

Most of all, consistent realism breaks down because it fails to provide any ground for purposive or meaningful action. If the sequence of cause and effect is sufficiently rigid to permit of the 'scientific prediction' of events, if our thought is irrevocably conditioned by our status and our interests, then both action and thought become devoid of purpose. If, as Schopenhauer maintains, 'the true philosophy of history consists of the insight that, throughout the jumble of all these ceaseless changes, we have ever before our eyes the same unchanging being, pursuing the same course to-day, yesterday and for ever',[8] then passive contemplation is all that remains to the individual. Such a conclusion is plainly repugnant to the most deep-seated belief of man about himself. That human affairs can be directed and modified by human action and human thought is a postulate so fundamental that its rejection seems scarcely compatible with existence as a human being. Nor is it in fact rejected by those realists who have left their mark on history. Machiavelli, when he exhorted his compatriots to be good Italians, clearly assumed that they were free to follow or ignore his advice. Marx, by birth and training a *bourgeois*, believed himself free to think and act like a proletarian, and regarded it as his mission to persuade others, whom he assumed to be equally free, to think and act likewise. Lenin, who wrote of the imminence of world revolution as a 'scientific prediction', admitted elsewhere that 'no situation exists from which there is absolutely no way out'.[9] In moments of crisis, Lenin appealed to his followers in terms which might equally well have been used by so thorough-going a believer in the power of the human will as Mussolini or by any other leader of any period: 'At the decisive moment and in the decisive place, you *must prove* the stronger, you must *be victorious*.'[10] Every realist, whatever his profession, is ultimately compelled to believe not only that there is something which man ought to think and do, but that there is something which he can think and do, and that his thought and action are neither mechanical nor meaningless.

We return therefore to the conclusion that any sound political thought must be based on elements of both utopia and reality. Where utopianism has become a hollow and intolerable sham, which serves merely as a disguise for the interests of the privileged, the realist performs an indispensable service in unmasking it. But pure realism can offer nothing but a naked struggle for power which makes any kind of international society impossible. Having demolished the current utopia with the weapons of realism, we still need to build a new utopia of our own, which will one day fall to the same weapons. The human will will continue to seek an escape from the logical consequences of realism in the vision of an international order which, as soon as it crystallizes itself into concrete political form, becomes tainted with self-interest and hypocrisy, and must once more be attacked with the instruments of realism.

Here, then, is the complexity, the fascination and the tragedy of all political life. Politics are made up of two elements – utopia and reality – belonging to two different planes which can never meet. There is no greater barrier to clear political thinking than failure to distinguish between ideals, which are utopia, and institutions, which are reality. The communist who set communism against democracy was usually thinking of communism as a pure ideal of equality and brotherhood, and of democracy as an institution which existed in Great Britain, France or the United States and which exhibited the vested interests, the inequalities and the oppression inherent in all political institutions. The democrat who made the same comparison was in fact comparing an ideal pattern of democracy laid up in heaven with communism as an institution existing in Soviet Russia with class divisions, its heresy hunts and its concentration camps. The comparison, made in each case between an ideal and an institution, is irrelevant and makes no sense. The ideal, once it is embodied in an institution, ceases to be an ideal and becomes the expression of a selfish interest, which must be destroyed in the name of a new ideal. This constant interaction of irreconcilable forces is the stuff of politics. Every political situation contains mutually incompatible elements of utopia and reality, of morality and power.

This point will emerge more clearly from the analysis of the nature of politics which we have now to undertake.

Notes

1. Treitschke, *Aufsätze*, iv. p. 428.
2. Engels, *Ludwig Feuerbach* (Engl. transl.), p. 23.
3. Marx and Engels, *Works* (Russian ed.), xv. p. 275.
4. Laski, *Communism*, p. 250.
5. R. Neibuhr, *Moral Man and Immoral Society*, p. 81.
6. Hitler, *Mein Kampf*, p. 231.
7. *Anti-Machiavel*, p. 248.
8. Schopenhauer, *Welt als Wille und Vorstellung*, ii. ch. 38.
9. Lenin, *Works* (2nd Russian ed.), xxv. p. 340.
10. Lenin, *Collected Works* (Engl. transl.), xxi. pt. i. p. 68.

Part Three

Politics, Power and Morality

CHAPTER SEVEN

The Nature of Politics

Man has always lived in groups. The smallest kind of human group, the family, has clearly been necessary for the maintenance of the species. But so far as is known, men have always from the most primitive times formed semi-permanent groups larger than the single family; and one of the functions of such a group has been to regulate relations between its members. Politics deals with the behaviour of men in such organized permanent or semi-permanent groups. All attempts to deduce the nature of society from the supposed behaviour of man in isolation are purely theoretical, since there is no reason to assume that such a man ever existed. Aristotle laid the foundation of all sound thinking about politics when he declared that man was by nature a political animal.

Man in society reacts to his fellow men in two opposite ways. Sometimes he displays egoism, or the will to assert himself at the expense of others. At other times he displays sociability, or the desire to co-operate with others, to enter into reciprocal relations of good will and friendship with them, and even to subordinate himself to them. In every society, these two qualities can be seen at work. No society can exist unless a substantial proportion of its members exhibits in some degree the desire for co-operation and mutual good will. But in every society some sanction is required to produce the measure of solidarity requisite for its maintenance; and this sanction is applied by a controlling group or individual acting in the name of the society. Membership of most societies is voluntary, and the only ultimate sanction which can be applied is expulsion. But the peculiarity of political society, which in the modern world takes the form of the state, is that membership is compulsory. The state, like other societies, must be based on some sense of common interests and obligations among its members. But coercion is regularly exercised by a governing group to enforce loyalty and obedience; and this coercion inevitably means that the governors control the governed and 'exploit' them for their own purposes.[1]

The dual character of political society is therefore strongly marked. Professor Laski tells us that 'every state is built upon the consciences of men'.[2] On the other hand, anthropology, as well as much recent history, teaches that 'war seems to be the main agency in producing the state';[3] and Professor Laski himself, in another passage, declares that 'our civilization is held together by fear rather than by good will'.[4] There is no contradiction between these apparently opposite views. When Tom Paine, in the *Rights of Man*, tries to confront Burke with the dilemma that 'governments arise either *out* of the people or *over* the people', the answer is that they do both. Coercion and conscience, enmity and good will, self-assertion and self-subordination, are present in every political society. The state is built up out of these two conflicting aspects of human nature. Utopia and reality, the ideal and the institution, morality and power, are from the outset inextricably blended in it. In the making of the United States, as a modern American writer has said, 'Hamilton stood for strength, wealth and power, Jefferson for the American dream'; and both the power and the dream were necessary ingredients.[5]

If this be correct, we can draw one important conclusion. The utopian who dreams that it is possible to eliminate self-assertion from politics and to base a political system on morality alone is just as wide of the mark as the realist who believes that altruism is an illusion and that all political action is based on self-seeking. These errors have both left their mark on popular terminology. The phrase 'power politics' is often used in an invidious sense, as if the element of power or self-assertion in politics were something abnormal and susceptible of elimination from a healthy political life. Conversely, there is a disposition, even among some writers who are not strictly speaking realists, to treat politics as the science of power and self-assertion and exclude from it by definition actions inspired by the moral consciousness. Professor Catlin describes the *homo politicus* as one who 'seeks to bring into conformity with his own will the wills of others, so that he may the better attain his own ends'.[6] Such terminological implications are misleading. Politics cannot be divorced from power. But the *homo politicus* who pursues nothing but power is as unreal a myth as the *homo economicus* who pursues nothing but gain. Political action must be based on a co-ordination of morality and power.

This truth is of practical as well as theoretical importance. It is as fatal in politics to ignore power as it is to ignore morality. The fate of China in the nineteenth century is an illustration of what happens to a country which is content to believe in the moral superiority of its own civilization and to despise the ways of power. The Liberal Government of Great Britain nearly came to grief in the spring of 1914 because it sought to pursue an Irish policy based on moral authority unsupported (or rather, directly opposed) by effective military power. In Germany, the Frankfort Assembly of 1848 is the classic example of the impotence of ideas divorced from power; and the Weimar Republic broke down because many of the policies it pursued – in fact, nearly all of them except its opposition to the communists – were unsupported, or actively opposed, by effective military power.[7] The utopian, who believes that democracy is not based on force, refuses to look these unwelcome facts in the face.

On the other hand, the realist, who believes that, if you look after the power, the moral authority will look after itself, is equally in error. The most recent form of this doctrine is embodied in the much-quoted phrase: 'The function of force is to give moral ideas times to take root'. Internationally, this argument was used in 1919 by those who, unable to defend the Versailles Treaty on moral grounds, maintained that this initial act of power would pave the way for subsequent moral appeasement. Experience has done little to confirm this comfortable belief. The same fallacy is implicit in the once popular view that the aim of British policy should be 'to rebuild the League of Nations, to make it capable of holding a political aggressor in restraint by armed power, and thereafter to labour faithfully for the mitigation of just and real grievances'.[8] Once the enemy has been crushed or the 'aggressor' restrained by force, the 'thereafter' fails to arrive. The illusion that priority can be given to power and that morality will follow, is just as dangerous as the illusion that priority can be given to moral authority and that power will follow.

Before proceeding, however, to consider the respective roles of power and morality in politics, we must take some note of the views of those who, though far from being realists, identify politics with power and believe that moral concepts must be altogether excluded from its scope. There is, according to this view, an essential antinomy between politics and morality; and the moral man as such will therefore have nothing to do with politics. This thesis has many attractions, and reappears at different periods of history and in different contexts. It takes at least three forms.

(i) Its simplest form is the doctrine of non-resistance. The moral man recognizes the existence of political power as an evil, but regards the use of power to resist power as a still greater evil. This is the basis of such doctrines of non-resistance as those of Jesus or of Gandhi, or of modern pacifism. It amounts, in brief, to a boycott of politics.

(ii) The second form of the antithesis between politics and morality is anarchism. The state, as the principal organ of political power, is 'the most flagrant, most cynical and most complete negation of humanity'.[9] The anarchist will use power to overthrow the state. This revolutionary power is, however, not thought of as political power, but as the spontaneous revolt of the outraged individual conscience. It does not seek to create a new political society to take the place of the old one, but a moral society from which power, and consequently politics, are completely eliminated. 'The principles of the Sermon on the Mount', an English divine recently remarked, would mean 'sudden death to civilised society'.[10] The anarchist sets out to destroy 'civilised society' in the name of the Sermon on the Mount.

(iii) A third school of thought starts from the same premise of the essential antithesis between morality and politics, but arrives at a totally different conclusion. The injunction of Jesus to 'render unto Caesar the things that are Caesar's, and unto God the things that are God's', implies the co-existence of two separate spheres: the political and the moral. But the moral man is under an obligation to assist – or at any rate not to obstruct – the politician in the discharge of his non-moral functions. 'Let every soul be subject to the higher powers. The powers that be are ordained of God.' We thus recognize politics as necessary but non-moral. This tradition, which remained dormant throughout the Middle Ages, when the ecclesiastical and the secular authority was theoretically one, was revived by Luther in order to effect his compromise between reformed church and state. Luther 'turned on the peasants of his day in holy horror when they attempted to transmute the "spiritual" kingdom into an "earthly" one by suggesting that the principles of the gospel had social significance.'[11] The division of functions between Caesar and God is implicit in the very conception of an 'established' church. But the tradition has been more persistent and more effective in Lutheran Germany than anywhere else. 'We do not consult Jesus', wrote a German liberal nineteenth-century pastor, 'when we are concerned with things which belong to the domain of the construction of the state and political economy';[12] and Bernhardi declared that 'Christian morality is personal and social, and in its nature cannot be political'.[13] The same attitude is inherent in the modern theology of Karl Barth, which insists that political and social evils are the necessary product of man's sinful nature and that human effort to eradicate them is therefore futile; and the doctrine that Christian morality has nothing to do with politics is vigorously upheld by the Nazi régime. This view is basically different from that of the realist who makes morality a function of politics. But in the field of politics it tends to become indistinguishable from realism.

The theory of the divorce between the spheres of politics and morality is superficially attractive, if only because it evades the insoluble problem of finding a moral justification for the use of force.[14] But it is not ultimately satisfying. Both non-resistance and anarchism are counsels of despair, which appear to find widespread acceptance only where men feel hopeless of achieving anything by political action; and the attempt to keep God and Caesar in watertight compartments runs too much athwart the deep-seated desire of the human mind to reduce its view of the world to some kind of moral order. We are not in the long run satisfied to believe that what is politically good is morally bad;[15] and since we can neither moralize power nor expel power from politics, we are faced with a dilemma which cannot be completely resolved. The planes of utopia and of reality never coincide. The ideal cannot be institutionalized, nor the institution idealized. 'Politics', writes Dr Niebuhr, 'will, to the end of history, be an area where conscience and power meet, where the ethical and coercive factors of human life will interpenetrate and work out their tentative and uneasy compromises.'[16] The compromises, like solutions of other human problems, will remain uneasy and tentative. But it is an essential part of any compromise that both factors shall be taken into account.

We have now therefore to analyse the part played in international politics by these two cardinal factors: power and morality.

Notes

1. 'Everywhere do I perceive a certain conspiracy of the rich men seeking their own advantage under the name and pretext of the commonwealth' (More, *Utopia*). 'The exploitation of one part of society by another is common to all past centuries' (*Communist Manifesto*).
2. *A Defence of Liberty against Tyrants* (*Vindiciae contra Tyrannos*), ed. Laski, Introd., p. 55.
3. Linton, *The Study of Man*, p. 240.
4. Laski, *A Grammar of Politics*, p. 20.
5. J. Truslow Adams, *The Epic of America*, p. 112. The idea that the state has a moral foundation in the consent of its citizens as well as a power foundation was propounded by Locke and Rousseau and popularized by the American and French revolutions. Two recent expressions of the idea may be quoted. The Czecho-Slovak declaration of independence of 18 October 1918 described Austria-Hungary as 'a state which has no justification for its existence, and which, since it refuses to accept the fundamental basis of modern world-organisation [i.e. self-determination], is only an artificial and unmoral construction'. In February 1938, Hitler told Schuschnigg, the then Austrian Chancellor, that 'a régime lacking every kind of legality and which in reality ruled only by force, must in the long run come into continually increasing conflict with public opinion' (speech in the Reichstag of 17 March 1938). Hitler maintained that the two pillars of the state are 'force' and 'popularity' (*Mein Kampf*, p. 579).
6. Catlin, *The Science and Method of Politics*, p. 309.
7. It is significant that the word *Realpolitik* was coined in the once famous treatise of von Rochau, *Grundsätze der Realpolitik* published in 1853, which was largely inspired by the lessons of Frankfort. The inspiration which Hiter's *Realpolitik* has derived from the lessons of the Weimar Republic is obvious.

8. Winston Churchill, *Arms and the Covenant*, p. 368. The argument that power is a necessary motive force for the remedy of 'just' grievances is further developed on pp. 209–16.

9. Bakunin, (*Oeuvres*, i. p. 150; cf. vi. p. 17: 'If there is a devil in all human history, it is this principle of command and authority.'

10. The Dean of St Paul's quoted in a leading article in *The Times*, 2 August 1937.

11. R. Niebuhr, *Moral Man and Immoral Society*, p. 77.

12. Quoted in W. F. Bruck, *Social and Economic History of Germany*, p. 65.

13. Bernhardi, *Germany and the Next War* (Engl. transl.), p. 29.

14. 'Force in the right place', as Mr Maxton once said in the House of Commons, is a meaningless conception, 'because the right place for me is exactly where I want to use it, and for him also, and for everyone else' (House of Commons, 7 November 1933: *Official Record*, col. 130). Force in politics is always the instrument of some kind of group interest.

15. Acton was fond of saying that 'great men are almost always bad men', and quotes Walpole's dictum that 'no great country was ever saved by good men' (*History of Freedom*, p. 219). Rosebery showed more acuteness when he remarked that 'there is one question which English people ask about great men: Was he "a good man"?' (*Napoleon: The Last Phase*, p. 364).

16. R. Niebuhr, *Moral Man and Immoral Society*, p. 4.

CHAPTER EIGHT

Power in International Politics

Politics are, then, in one sense always power politics. Common usage applies the term 'political' not to all activities of the state, but to issues involving a conflict of power. Once this conflict has been resolved, the issue ceases to be 'political' and becomes a matter of administrative routine. Nor is all business transacted between states 'political'. When states cooperate with one another to maintain postal or transport services, or to prevent the spread of epidemics or suppress the traffic in drugs, these activities are described as 'non-political' or 'technical'. But as soon as an issue arises which involves, or is thought to involve, the power of one state in relation to another, the matter at once becomes 'political'. While politics cannot be satisfactorily defined exclusively in terms of power, it is safe to say that power is always an essential element of politics. In order to understand a political issue, it is not enough (as it would be in the case of a technical or a legal issue) to know what the point at issue is. It is necessary also to know between whom it has arisen. An issue raised by a small number of isolated individuals is not the same political fact as the same issue raised by a powerful and well-organized trade union. A political issue arising between Great Britain and Japan is something quite different from what may be formally the same issue between Great Britain and Nicaragua. 'Politics begin where the masses are,' said Lenin, 'not where there are thousands, but where there are millions, that is where serious politics begin.'[1]

There have been periods of history when it might have been superfluous to dwell on this obvious fact, and when Engels' dictum that 'without force and iron ruthlessness nothing is achieved in history'[2] would have passed as a platitude. But in the comparatively well-ordered world of nineteenth-century liberalism, subtler forms of compulsion successfully concealed from the unsophisticated the continuous but silent workings of political power; and in democracies, at any rate, this concealment is still partially effective.[3] After the first world war, the liberal tradition was carried into international politics. Utopian writers from the English-speaking countries seriously believed that the establishment of the League of Nations meant the elimination of power from international relations, and the substitution of discussion for armies and navies. 'Power politics' were regarded as a mark of the bad old times, and became a term of abuse. That this belief should have persisted for more than ten years was due to the circumstance that the Great Powers whose main interest was the preservation of the *status quo* enjoyed throughout that time a virtual monopoly of power. A game of chess between a world champion and a schoolboy would be so rapidly and so effortlessly won that the innocent onlooker might be pardoned for assuming that little skill was necessary to play chess. In the same way, the simple-minded spectator of the game of international politics could assume, between 1920 and 1931, that power played little part in the game. What was commonly called the 'return to power politics' in 1931 was, in fact, the termination of the monopoly of power enjoyed by the *status quo* Powers. Stalin's lament that '*in our days* it is not the custom to reckon with the weak', and Neville Chamberlain's remark that '*in the world as we find it to-day* an unarmed nation has little chance of making its voice heard',[4] were curious tributes – more surprising in the professed Marxist than in the inheritor of a British nineteenth-century tradition – to the illusion that there was once a time when weak and unarmed countries played an effective role in international politics.

The assumption of the elimination of power from politics could only result from a wholly uncritical attitude towards political problems. In the affairs of the League of Nations, formal equality and the participation of all in debate did not render the power factor any less decisive. The founders of the League themselves entertained no such illusion. House originally thought that only Great Powers should be admitted to the League at all.[5] In the earliest British and American drafts of the Covenant, it was contemplated that membership of the Council of the League would be limited to Great Powers; and Lord Cecil noted on one of these drafts that 'the smaller Powers would in any case not exercise any considerable influence'.[6] This prevision was fulfilled. An Italian delegate testified that during the long period of his regular attendances at Geneva he 'never saw a dispute of any importance settled otherwise than by an agreement between the Great Powers', and that the procedure of the League was 'a system of detours, all of which lead to one or other of these two issues: agreement or disagreement between Great Britain, Italy, France and Germany'.[7] 'Despite our juridical equality here,' said Mr De Valera a little later, 'in matters such as European peace the small states are powerless.'[8] The decisions on the application of sanctions against Italy in the winter of 1935–36 were, in effect, taken solely by Great Britain and France, the possessors of effective military and economic power in the Mediterranean. The minor Powers followed their lead; and one of them was actually 'compensated' by Great Britain and France for so doing.

Nor was it only at Geneva that the weak Powers set their course to match that of the strong. When Great Britain took her currency off the gold standard in September 1931, several minor Powers were obliged to follow her example. When France abandoned the gold standard in September 1936, Switzerland and Holland – the last free gold countries – were compelled to follow suit, and several other smaller countries had to alter the value of their currencies. When France was militarily supreme in Europe in the nineteen-twenties, a number of smaller Powers grouped themselves under her aegis. When German military strength eclipsed that of France, most of these Powers made declarations of neutrality or veered to the side of Germany. The alleged 'dictatorship of the Great Powers', which is sometimes denounced by utopian writers as if it were a wicked policy deliberately adopted by certain states, is a fact which constitutes something like a 'law of nature' in international politics.

It is necessary at this point to dispel the current illusion that the policy of those states which are, broadly speaking, satisfied with the *status quo* and whose watchword is 'security', is somehow less concerned with power than the policy of the dissatisfied states, and that the popular phrase 'power politics' applies to the acts of the latter but not to those of the former. This illusion, which has an almost irresistible attraction for the publicists of the satisfied Powers, is responsible for much confused thinking about international politics. The pursuit of 'security' by satisfied Powers has often been the motive of flagrant examples of power politics. In order to secure themselves against the revenge of a defeated enemy, victorious Powers have in the past resorted to such measures as the taking of hostages, the mutilation or enslavement of males of military age or, in modern times, the dismemberment and occupation of territory or forced disarmament. It is profoundly misleading to represent the struggle between satisfied and dissatisfied Powers as a struggle between morality on one side and power on the other. It is a clash in which, whatever the moral issue, power politics are equally predominant on both sides.

The history of the Locarno Treaty is a simple and revealing illustration of the working of power politics. The first proposal for a treaty guaranteeing Germany's western frontier was made by Germany in December 1922, and was emphatically rejected by Poincaré. At this period (it was the eve of the Ruhr invasion), Germany had everything to fear from France, and France nothing to fear from a helpless Germany; and the treaty had no attraction for France. Two years later the position had changed. The Ruhr invasion had brought little profit to France, and had left her perplexed as to the next step. Germany might one day be powerful again. Germany, on the other hand, still feared the military supremacy of France, and hankered after a guarantee. It was the psychological moment when French fear of Germany was about equally balanced by Germany's fear of France; and a treaty which had not been possible two years before, and would not have been possible five years later, was now welcome to both. Moreover, the power interests of Great Britain coincided with those of Germany. Germany had abandoned hope of securing a revision of her western, but not of her other, frontiers. Great Britain was prepared to guarantee Germany's western, but not her other, frontiers. Germany, anxious to expedite the withdrawal of the Allied army from the Rhineland, had as yet no hope of breaking down the restrictions imposed by the demilitarization clauses of the Versailles Treaty; and she was therefore quite prepared to purchase the new agreement by reaffirming her acceptance of those clauses and placing them under a guarantee.

Such was the background of the famous Locarno Treaty. Its success was a striking one. For years afterwards, attempts were made to repeat it in other fields. Mediterranean and Eastern European 'Locarnos' were canvassed; and their failure to materialize disappointed and puzzled people who believed that international problems everywhere could be solved by devices of the same standard pattern, and who failed to understand that the Locarno Treaty was an expression of the power politics of a particular period and locality. Ten years after its conclusion, the delicate balance on which it rested had disappeared. France feared Germany more than ever. But Germany no longer feared anything from France. The treaty no longer had any meaning for Germany save as an affirmation of the demilitarization clauses of the Versailles Treaty which she could now hope to overthrow. The only part of the Locarno Treaty which still corresponded to the situation of power politics was the British guarantee to France and Belgium. This was repeated by Great Britain after the rest of the treaty had been denounced by Germany. The history of Locarno is a classic instance of power politics. It remains incomprehensible to those who seek uniform *a priori* solutions of the problem of security, and regard power politics as an abnormal phenomenon visible only in periods of crisis.

Failure to recognize that power is an essential element of politics has hitherto vitiated all attempts to establish international forms of government, and confused nearly every attempt to discuss the subject. Power is an indispensable instrument of government. To internationalize government in any real sense means to internationalize power; and international government is, in effect, government by that state which supplies the power necessary for the purpose of governing. The international governments set up by the Versailles Treaty in various parts of Europe were temporary in character, and had not therefore to face the problems of a long-term policy. But even these illustrate the intimate connexion between government and power. The Inter-Allied High Commission, which exercised in the occupied Rhineland such functions of government as were necessary for the security of the Allied troops, worked smoothly so long as British and French policies coincided. When the Ruhr crisis caused a serious difference of opinion between the British and French Governments, French policy was applied in the zones occupied by French and Belgian troops and British policy in the zone occupied by British troops, the policy of the government being determined by the nationality of the power on which it rested. The Inter-Allied Commission appointed to conduct the plebiscite in Upper Silesia pursued the French policy of favouring Poland so long as the Allied troops on which its authority depended were supplied almost exclusively by France. This policy was corrected only when British troops were sent to the area. The effective control of any government depends on the source of its power.

The problem of international government and power was raised in a more acute form by the mandates system and by the proposal frequently put forward that the government of some or all colonial territories shall be 'internationalized'. We are here faced by an issue of permanent government, involving the formulation of long-term policy, and different in kind from that of temporary international collaboration between allies under stress of war or for the purpose of implementing a treaty jointly imposed. Its nature may be illustrated from the case of Palestine. Policy in Palestine was dependent on the amount of military force available for use there, and had therefore to be determined not by the Mandates Commission, which had no power at its disposal, but by the British Government, which supplies the power; for whatever view might be taken by the Mandates Commission, it was unthinkable that British troops could be used to carry out a policy of which the British Government or the British electorate did not approve.[9] Under any international system of government, policy would depend, at critical moments, on the decision of the state supplying the power on which the authority of the government depended. If, as would almost inevitably happen, the control of an international territory were divided geographically among the forces of different states, the different zones would, in periods of international discord, pursue discordant policies: and the old international rivalries would recur in a new and equally dangerous form. Problems of economic development would be not less baffling. The international administration of colonial areas, wrote Lugard, himself an experienced and enlightened administrator, 'would paralyse all initiative by the dead hand of a super-bureaucracy devoid of national sentiment and stifling to all patriotism, and would be very disadvantageous to the countries concerned'.[10] Any real international government is impossible so long as power, which is an essential condition of government, is organized nationally. The international secretariat of the League of Nations was able to function precisely because it was a non-political civil service, had no responsibility for policy, and was therefore independent of power.

Political power in the international sphere may be divided, for purposes of discussion, into three categories: (a) military power, (b) economic power, (c) power over opinion. We shall find, however, that these categories are closely interdependent; and though they are theoretically separable, it is difficult in practice to imagine a country for any length of time possessing one kind of power in isolation from the others. In its essence, power is an indivisible whole. 'The laws of social dynamics', a recent critic has said, 'are laws which can only be stated in terms of power, not in terms of this or that form of power.'[11]

(a) Military power

The supreme importance of the military instrument lies in the fact that the *ultima ratio* of power in international relations is war. Every act of the state, in its power aspect, is directed to war, not as a desirable weapon, but as a weapon which it may require in the last resort to use. Clausewitz's famous aphorism that 'war is nothing but the continuation of political relations by other means' has been repeated with approval both by Lenin and by the Communist International;[12] and Hitler meant much the same thing when he said that 'an alliance whose object does not include the intention to fight is meaningless and useless'.[13] In the same sense, Mr Hawtrey defines diplomacy as 'potential war'.[14] These are half-truths. But the important thing is to recognize that they are true. War lurks in the background of international politics just as revolution lurks in the background of domestic politics. There are few European countries where, at some time during the past thirty years, potential revolution has not been an important factor in politics;[15] and the international community has in this respect the closest analogy to those states where the possibility of revolution is most frequently and most conspicuously present to the mind.

Potential war being thus a dominant factor in international politics, military strength becomes a recognized standard of political values. Every great civilization of the past has enjoyed in its day a superiority of military power. The Greek city-state rose to greatness when its hoplite armies proved more than a match for the Persian hordes. In the modern world, Powers (the word itself is significant enough) are graded according to the quality and the supposed efficiency of the military equipment, including manpower, at their disposal. Recognition as a Great Power is normally the reward of fighting a successful large-scale war. Germany after the Franco-Prussian War, the United States after the war with Spain, and Japan after the Russo-Japanese War are familiar recent instances. The faint doubt attaching to Italy's status as a Great Power is partly due to the fact that she has never proved her prowess in a first-class war. Any symptom of military inefficiency or unpreparedness in a Great Power is promptly reflected in its political status. The naval mutiny at Invergordon in September 1931 was the final blow to British prestige which compelled Great Britain to devalue her currency. The execution of the leading Soviet generals for alleged treason in June 1937 was thought to reveal so much weakness in the Soviet military machine that the political influence of Soviet Russia suffered a sudden and severe slump. Statesmen of all the Great Powers periodically make speeches extolling the efficiency of their armies, navies and air forces; and military parades and reviews are organized in order to impress the world with the military strength and consequent political standing of the nation. In international crises, fleets, troops or air squadrons show themselves conspicuously at crucial points for the same purpose.

These facts point to the moral that foreign policy never can, or never should, be divorced from strategy. The foreign policy of a country is limited not only by its aims, but also by its military strength or, more accurately, by the ratio of its military strength to that of other countries. The most serious problem involved in the democratic control of foreign policy is that no government can afford to divulge full and frank information about its own military strength, or all the knowledge it possesses about the military strength of other countries. Public discussions of foreign policy are therefore conducted in partial or total ignorance of one of the factors which must be decisive in determining it. A constitutional rule of long standing precludes private members of Parliament from proposing motions which entail public expenditure. The same restraint might well be exercised in advocating policies which entail risk of war; for only the government and its advisers can assess the chances with anything like complete knowledge of the relevant facts. Many contemporary books and speeches about international politics are reminiscent of those ingenious mathematical problems which the student is invited to solve by ignoring the weight of the elephant. The solutions proposed are neat and accurate on the abstract plane, but are obtained by leaving out of account the vital strategic factor. Even so important, and in many ways so valuable, a work as the annual *Survey of International Affairs* frequently soars into the realms of fancy when it embarks on criticism of policy, precisely because it neglects those military limitations which are always present to the minds of those who have to solve problems of foreign policy in real life. If every prospective writer on international affairs in the last twenty years had taken a compulsory course in elementary strategy, reams of nonsense would have remained unwritten.

Military power, being an essential element in the life of the state, becomes not only an instrument, but an end in itself. Few of the important wars of the last hundred years seem to have been waged for the deliberate and conscious purpose of increasing either trade or territory. The most serious wars are fought in order to make one's own country militarily stronger or, more often, to prevent another country from becoming militarily stronger, so that there is much justification for the epigram that 'the principal cause of war is war itself'.[16] Every stage of the Napoleonic Wars was devised to prepare the way for the next stage: the invasion of Russia was undertaken in order to make Napoleon strong enough to defeat Great Britain. The Crimean War was waged by Great Britain and France in order to prevent Russia from becoming strong enough to attack their Near Eastern possessions and interests at some future time. The origin of the Russo-Japanese War of 1904–5 is described as follows in a note addressed to the League of Nations by the Soviet Government in 1924: 'When the Japanese torpedo-boats attacked the Russian fleet at Port Arthur in 1904, it was clearly an act of aggression from a technical point of view, but, politically speaking, it was an act caused by the aggressive policy of the Tsarist Government towards Japan, who, in order to forestall the danger, struck the first blow at her adversary.'[17] In 1914, Austria sent an ultimatum to Serbia because she believed that Serbians were planning the downfall of the Dual Monarchy; Russia feared that Austria-Hungary, if she defeated Serbia, would be strong enough to menace her; Germany feared that Russia, if she defeated Austria-Hungary, would be strong enough to menace her; France had long believed that Germany, if she defeated Russia, would be strong enough to menace her, and had therefore concluded the Franco-Russian alliance; and Great Britain feared that Germany, if she defeated France and occupied Belgium, would be strong enough to menace her. Finally, the United States came to fear that Germany, if she won the war, would be strong enough to menace them. Thus the war, in the minds of all the principal combatants, had a defensive or preventive character. They fought in order that they might not find themselves in a more unfavourable position in some future war. Even colonial acquisitions have often been prompted by the same motive. The consolidation and formal annexation of the original British settlements in Australia were inspired by fear of Napoleon's alleged design to establish French colonies there. Military, rather than economic, reasons dictated the capture of German colonies during the war of 1914 and afterwards precluded their return to Germany.

It is perhaps for this reason that the exercise of power always appears to beget the appetite for more power. There is, as Dr Niebuhr says, 'no possibility of drawing a sharp line between the will-to-live and the will-to-power'.[18] Nationalism, having attained its first objective in the form of national unity and independence, develops almost automatically into imperialism. International politics amply confirm the aphorisms of Machiavelli that 'men never appear to themselves to possess securely what they have unless they acquire something further from another',[19] and of Hobbes that man 'cannot assure the power and means to live well which he hath present, without the acquisition of more'.[20] Wars, begun for motives of security, quickly become wars of aggression and self-seeking. President McKinley invited the United States to intervene in Cuba against Spain in order 'to secure a full and final termination of hostilities between the Government of Spain and the people of Cuba and to secure on the island the establishment of a stable government'.[21] But by the time the war was over the temptation to self-aggrandizement by the annexation of the Philippines had become irresistible. Nearly every country participating in the first world war regarded it initially as a war of self-defence; and this belief was particularly strong on the Allied side. Yet during the course of the war, every Allied Government in Europe announced war aims which included the acquisition of territory from the enemy Powers. In modern conditions, wars of limited objective have become almost as impossible as wars of limited liability. It is one of the fallacies of the theory of collective security that war can be waged for the specific and disinterested purpose of 'resisting aggression'. Had the League of Nations in the autumn of 1935, under the leadership of Great Britain, embarked on 'military sanctions' against Italy, it would have been impossible to restrict the campaign to the expulsion of Italian troops from Abyssinia. Operations would in all probability have led to the occupation of Italy's East African colonies by Great Britain and France, of Trieste, Fiume and Albania by Yugoslavia, and of the islands of the Dodecanese by Greece or Turkey or both; and war aims would have been announced, precluding on various specious grounds the restoration of these territories to Italy. Territorial ambitions are just as likely to be the product as the cause of war.

(b) Economic power

Economic strength has always been an instrument of political power, if only through its association with the military instrument. Only the most primitive kinds of warfare are altogether independent of the economic factor. The wealthiest prince or the wealthiest city-state could hire the largest and most efficient army of mercenaries; and every government was therefore compelled to pursue a policy designed to further the acquisition of wealth. The whole progress of civilization has been so closely bound up with economic development that we are not surprised to trace, throughout modern history, an increasingly intimate association between military and economic power. In the prolonged conflicts which marked the close of the Middle Ages in Western Europe, the merchants of the towns, relying on organized economic power, defeated the feudal barons, who put their trust in individual military prowess. The rise of modern nations has everywhere been marked by the emergence of a new middle class economically based on industry and trade. Trade and finance were the foundation of the shortlived political supremacy of the Italian cities of the Renaissance and later of the Dutch. The principal international wars of the period from the Renaissance to the middle of the eighteenth century were trade wars (some of them were actually so named). Throughout this period, it was universally held that, since wealth is a source of political power, the state should seek actively to promote the acquisition of wealth; and it was believed that the right way to make a country powerful was to stimulate production at home, to buy as little as possible from abroad, and to accumulate wealth in the convenient form of precious metals. Those who argued in this way afterwards came to be known as mercantilists. Mercantilism was a system of economic policy based on the hitherto unquestioned assumption that to promote the acquisition of wealth was part of the normal function of the state.

The separation of economics from politics

The *laissez-faire* doctrine of the classical economists made a frontal attack on this assumption. The principal implications of *laissez-faire* have already been discussed. Its significance in the present context is that it brought about a complete theoretical divorce between economics and politics. The classical economists conceived a natural economic order with laws of its own, independent of politics and functioning to the greatest profit of all concerned when political authority interfered least in its automatic operation. This doctrine dominated the economic thought, and to some extent the economic practice (though far more in Great Britain than elsewhere), of the nineteenth century. The theory of the nineteenth-century liberal state presupposed the existence side by side of two separate systems. The political system, which was the sphere of government, was concerned with the maintenance of law and order and the provision of certain essential services, and was thought of mainly as a necessary evil. The economic system, which was the preserve of private enterprise, catered for the material wants and, in doing so, organized the everyday lives of the great mass of the citizens.[22] In current English theory, the doctrine of the separation of politics and economics was sometimes carried to astonishing lengths. 'Is it true', asked Sir Normal Angell shortly before the first world war, 'that wealth and prosperity and well-being depend on the political power of nations, or indeed that one has anything whatever to do with the other?'[23] And the whole argument depends on the confident assumption that every intelligent reader will answer in the negative. As late as 1915, an English philosopher detected 'an ineradicable tendency that, as wealth and its control and enjoyment go to the productive class, so power and prestige go to the professional class', and regarded this separation of economic from political power as not only ineradicable but 'essential to a decent society'.[24]

Even before 1900, a more penetrating analysis might have shown that the illusion of a divorce between politics and economics was fast breaking down. It is still open to debate whether late nineteenth-century imperialism should be regarded as an economic movement using political weapons, or as a political movement using economic weapons. But that economics and politics marched hand in hand towards the same objective is clear enough. 'Is it not precisely the hallmark of British statesmanship', asked Hitler, 'to draw economic advantages from political strength, and to transform every economic gain back into political power?'[25] The first world war, by overtly reuniting economics and politics, in both domestic and foreign policy, hastened a development which was already on the way. It was now revealed that the nineteenth century, while purporting to take economics altogether out of the political sphere, had in fact forged economic weapons of unparalleled strength for use in the interests of national policy. A German staff officer had remarked to Engels in the 1880s that 'the basis of warfare is primarily the general economic life of peoples';[26] and this diagnosis was amply confirmed by the experiences of 1914–18. In no previous war had the economic life of belligerent nations been so completely and ruthlessly organized by the political authority. In the age-long alliance between the military and the economic arm, the economic arm for the first time was an equal, if not a superior, partner. To cripple the economic system of an enemy Power was as much a war aim as to defeat his armies and fleets. 'Planned economy', which means the control by the state for political purposes of the economic life of the nation, was a development of the first world war.[27] 'War potential' has become another name for economic power.

We have now therefore returned, after the important, but abnormal, *laissez-faire* interlude of the nineteenth century, to the position where economics can be frankly recognized as a part of politics. We can thus resolve the controversy, which is in large part a product of nineteenth-century ideas and terminology, about the so-called economic interpretation of history. Marx was overwhelmingly right when he insisted on the increasing importance of the role played by economic forces in politics; and since Marx, history can never be written again exactly as it was written before him. But Marx believed, just as firmly as did the *laissez-faire* liberal, in an economic system with laws of its own working independently of the state, which was its adjunct and its instrument. In writing as if economics and politics were separate domains, one subordinate to the other, Marx was dominated by nineteenth-century presuppositions in much the same way as his more recent opponents who are equally sure that 'the primary laws of history are political laws, economic laws are secondary'.[28] Economic forces are in fact political forces. Economics can be treated neither as a minor accessory of history, nor as an independent science in the light of which history can be interpreted. Much confusion would be saved by a general return to the term 'political economy', which was given the new science by Adam Smith himself and not abandoned in favour of the abstract 'economics', even in Great Britain itself, till the closing years of the nineteenth century.[29] The science of economics presupposes a given political order, and cannot be profitably studied in isolation from politics.

Some fallacies of the separation of economics from politics

It would have been unnecessary to dwell at length on this point if its importance had been either purely historical or purely theoretical. The illusion of a separation between politics and economics – a belated legacy of the *laissez-faire* nineteenth century – had ceased to correspond to any aspect of current reality. But it continued to persist in thought about international politics, where it created no little confusion. An immense amount of discussion was devoted to the meaningless question whether (as the Economic Conference of 1927 supposed)[30] our political troubles have economic causes or whether (as the Van Zeeland report suggested)[31] our economic troubles have political causes, and to the equally meaningless conundrum whether the problem of raw materials is political or economic. Similar confusion was produced by the declaration of the British Government in 1922 that the rate of Jewish immigration into Palestine would be determined by 'the economic capacity of the country', supplemented in 1931 by the further statement that 'the considerations relevant to the limits of absorptive capacity are purely economic considerations'. It was not until 1937 that a Royal Commission discovered that 'since Arabs are hostile to Jewish immigration, the factor of "hostility between the two peoples" necessarily assumes immediate economic importance'.[32] Indeed every issue of migration and refugees has been complicated by the supposition that there is some objective economic test of absorptive capacity. The conflict between two opposite and equally defensible interpretations of the promise in the Treaty of Neuilly 'to ensure the economic outlets of Bulgaria to the Aegean Sea' was another instance of confusion arising from the too light-hearted use of this elusive word. Attempts to solve international problems by the application of economic principles divorced from politics are doomed to sterility.

The most conspicuous practical failure caused by the persistence of this nineteenth-century illusion was the breakdown of League sanctions in 1936. Careful reading of the text of Article 16 of the Covenant acquits its framers of responsibility for the mistake. Paragraph 1 prescribes the economic weapons, paragraph 2 the military weapons, to be employed against the violator of the Covenant. Paragraph 2 is clearly complementary to paragraph 1, and assumes as a matter of course that, in the event of an application of sanctions, 'armed forces' would be required 'to protect the Covenants of the League'. The only difference between the two paragraphs is that, whereas all members of the League would have to apply the economic weapons, it would be natural to draw the necessary armed forces from those members which possessed them in sufficient strength and in reasonable geographical proximity to the offender.[33] Subsequent commentators, obsessed with the assumption that economics and politics were separate and separable things, evolved the doctrine that paragraphs 1 and 2 of Article 16 were not complementary, but alternative, the difference being that 'economic sanctions' were obligatory and 'military sanctions' optional. This doctrine was eagerly seized on by the many who felt that the League might conceivably be worth a few million pounds' worth of trade, but not a few million human lives; and in the famous 1934 Peace Ballot in Great Britain, some two million deluded voters expressed simultaneously their approval of economic, and their disapproval of military, sanctions. 'One of the many conclusions to which I have been drawn', said Lord Baldwin at this time, 'is that there is no such thing as a sanction which will work, which does not mean war.'[34] But the bitter lesson of 1935–36 was needed to drive home the truth that in sanctions, as in war, the only motto is 'all or nothing', and that economic power is impotent if the military weapon is not held in readiness to support it.[35] Power is indivisible; and the military and economic weapons are merely different instruments of power.[36]

A different, and equally serious, form in which this illusory separation of politics and economics can be traced is the popular phraseology which distinguishes between 'power' and 'welfare', between 'guns' and 'butter'. 'Welfare arguments are "economic",' remarks an American writer, 'power arguments are "political".'[37] This fallacy is particularly difficult to expose because it appears to be deducible from a familiar fact. Every modern government and every parliament is continually faced with the dilemma of spending money on armaments or social services; and this encourages the illusion that the choice really lies between 'power' and 'welfare', between political guns and economic butter. Reflexion shows, however, that this is not the case. The question asked never takes the form, Do you prefer guns or butter? For everyone (except a handful of pacifists in those Anglo-Saxon countries which have inherited a long tradition of uncontested security) agrees that, in case of need, guns must come before butter. The question asked is always either, Have we already sufficient guns to enable us to afford some butter? or, Granted that we need x guns, can we increase revenue sufficiently to afford more butter as well? But the neatest exposure of this fallacy comes from the pen of Professor Zimmern; and the exposure is none the less effective for being unconscious. Having divided existing states on popular lines into those which pursue 'welfare' and those which pursue 'power', Professor Zimmern revealingly adds that 'the welfare states, taken together, enjoy a preponderance of power and resources over the power states',[38] thereby leading us infallibly to the correct conclusion that 'welfare states' are states which, already enjoying a preponderance of power, are not primarily concerned to increase it, and can therefore afford butter, and 'power states' those which, being inferior in power, are primarily concerned to increase it, and devote the major part of their resources to this end. In this popular terminology, 'welfare states' are those which possess preponderant power, and 'power states' those which do not. Nor is this classification as illogical as it may seem. Every Great Power takes the view that the minimum number of guns necessary to assert the degree of power which it considers requisite takes precedence over butter, and that it can only pursue 'welfare' when this minimum has been achieved. For many years prior to 1933, Great Britain, being satisfied with her power, was a 'welfare state'. After 1935, feeling her power contested and inadequate, she became a 'power state'; and even the Opposition ceased to press with any insistence the prior claim of the social services. The contrast is not one between 'power' and 'welfare', and still less between 'politics' and 'economics', but between different degrees of power. In the pursuit of power, military and economic instruments will both be used.

Autarky

Having thus established that economics must properly be regarded as an aspect of politics, we may divide into two broad categories the methods by which economic power is pressed into the service of national policy. The first will contain those measures whose purpose is defined by the convenient word autarky; the second, economic measures directly designed to strengthen the national influence over other countries.

Autarky, or self-sufficiency, was one of the aims of the mercantilist policy, and has indeed been pursued by states from the earliest times. But the problem of autarky is nevertheless distinctively modern. In the Middle Ages, autarky was a natural and necessary condition of economic life; for the long-distance transport of any goods other than those of small bulk and great value was unremunerative. From the close of the Middle Ages, transport gradually became safer, cheaper and more rapid. Countries became less completely self-dependent; and a rising standard of life was based in part on the international exchange of specialized products. But it is only within the last hundred years that the coming of steam has made transport by land and sea so rapid and cheap that the cost of transport of most commodities is now insignificant in relation to the cost of production, and it is in many cases immaterial whether an article is produced 500 or 5000 miles from the point where it will be used or consumed. Mass-production methods, under which commodities become cheaper the more of them are produced in the same place, have further promoted concentration. Not only are our needs today more highly specialized than ever before, but we live in a world where, for the first time in history, it might, from the standpoint of cost, be possible – and perhaps even desirable – to grow all the wheat consumed by the human race in Canada, and all the wool in Australia, to manufacture all the motor cars in Detroit and all the cotton clothing in England or Japan. Internationally, the consequences of absolute *laissez-faire* are as fantastic and as unacceptable as are the consequences of *laissez-faire* within the state. In modern conditions the artificial promotion of some degree of autarky is a necessary condition of orderly social existence.

Autarky is, however, not only a social necessity, but an instrument of political power. It is primarily a form of preparedness for war. In the mercantilist period, it was commonly asserted, both in Britain and elsewhere, that the military power of the state depended on the production of manufactured goods. Adam Smith made his famous exceptions to the doctrine of *laissez-faire* when he approved of the British Navigation Act and the bounties on British sail-cloth and British gunpowder. But the principle of autarky received its classic definition from the pen of Alexander Hamilton, who in 1791, being then Secretary of the United States Treasury, made a report to the House of Representatives which enunciates, in words which might have been written to-day, the whole modern doctrine of autarky. Hamilton had been instructed to advise on 'the means of promoting such [manufactures] as will tend to render the United States independent of foreign nations for military and other essential supplies'. One short passage may be quoted from the report:

Not only the wealth but the independence and security of a country appear to be materially connected with the prosperity of manufactures. Every nation, with a view to these great objects, ought to endeavour to possess within itself all the essentials of national supply. ... The extreme embarrassments of the United States during the late war, from an incapacity of supplying themselves, are still a matter of keen recollection; a future war might be expected to exemplify the mischief and dangers of a situation to which that capacity is still, in too great a degree, applicable, unless changed by timely and vigorous action.

And Hamilton went on to examine in turn all the methods by which the desired result might be attained – duties, prohibitions, bounties and premiums.[39] In Germany, just fifty years later, List argued that 'on the development of the German protective system depend the existence, the independence and the future of the German nationality';[40] and in the latter half of the nineteenth century successive Prussian victories drove home the intimate connexion between a highly developed industrial system and military power.

Throughout this period Great Britain, in virtue of her industrial supremacy, enjoyed virtually complete autarky in all industrial products, though not in the raw materials required to produce them. In food supplies, she ceased to be self-supporting about 1830. But this defect was in large part remedied by her naval power, the maintenance of which became one of her chief preoccupations. A Royal Commission on the Supply of Food and Raw Materials in Time of War, which reported in 1905, discussed, but rejected, plans for the precautionary storage in Great Britain of reserve supplies, and did not even discuss any plan for encouraging home production. Complete reliance was placed on the capacity of the navy to protect the ordinary channels of trade, and thereby make up for the inevitable absence of sufficient supplies at home.[41] The now current view that nineteenth-century statesmen were not alive to the political desirability of autarky, or of some adequate substitute for it, is not borne out by facts.

The effect of the first world war on the whole concept of economics has already been discussed. The impulse which it gave to the pursuit of autarky was immediate and powerful. Blockade, and the diversion of a large part of the world's shipping to the transport of troops and munitions, imposed more or less stringent measures of autarky on both belligerents and neutrals. For four years, the Central Powers were compelled to depend exclusively on their own resources, and to realize in spite of themselves Fichte's ideal of *The Closed Commercial State*. Even for the Allied Powers, the new weapon of the submarine made reliance on overseas imports as an alternative to autarky more precarious than it had hitherto been supposed. Nor did the Allied Governments, at any rate, appear to regard autarky as a regrettable and temporary expedient. In June 1916, they met in Paris to discuss postwar economic policy, and decided 'to take the necessary steps without delay to render themselves independent of the enemy countries in so far as regards the raw materials and manufactured articles essential to their normal economic activities'.[42] In the following year, a British Royal Commission drew up a list of articles in respect of which it had been established 'that the possibility of economic pressure from foreign countries controlling supplies of raw materials requires especially to be guarded against, and that government action is most needed in order to promote economic independence'; and this policy was carried into effect in the Safeguarding of Industries Act of 1921. Where home supplies were not available, the unfettered control of overseas supplies became a primary objective. The desire to control adequate supplies of oil inspired an active British policy in more than one oil-producing country.

Internationally, the important part played by the blockade in winning the war made inevitable the prominence of 'economic sanctions' in the constitution of the League of Nations. It was clear that blockade was likely to be applied more vigorously than ever in another war; and autarky was developed as the natural defensive armament against the weapon of blockade. The actual use of this weapon against Italy in 1935 added point to the moral. 'November 18, 1935, marks the starting point of a new chapter in Italian history,' said Mussolini to the National Guild Assembly on 23 March 1936. '... The new phase of Italian history will be determined by this postulate: to secure within the briefest time possible the greatest possible measure of economic independence.' There was, in fact, little novelty in this doctrine, which was merely a paraphrase of what had been said by Hamilton, by List, and by the British Royal Commission of 1917. But the growing international tension threw the problem into sharp relief. A well-known American publicist urged the joint buying by Great Britain and the United States of 'metals of strategic importance' with the object of 'removing the great bulk of these important metals from the markers in which the dictatorial and "have-not" Powers must buy them'.[43] 'No measure', added a British writer, 'would do more to weaken a German rearmament programme than a British decision to purchase the entire available output of Swedish ore.'[44] It scarcely required such warnings to convince governments of the military value of autarky.

The development of synthetic materials by Germany and the accumulation by Great Britain of stocks of foodstuffs and essential raw materials were two of many significant symptoms. Autarky, like other elements of power, is expensive. It may cost a country as much to make itself self-supporting in some important commodity as to build a battleship. The expenditure may turn out to be wasteful, and the acquisition not worth the cost. But to deny that autarky is an element of power, and as such desirable, is to obscure the issue.

Economic power as an instrument of policy

The second use of the economic weapon as an instrument of national policy, i.e. its use to acquire power and influence abroad, has been so fully recognized and freely discussed that the briefest summary will suffice here. It takes two principal forms: (*a*) the export of capital, and (*b*) the control of foreign markets.

(*a*) The export of capital has in recent times been a familiar practice of powerful states. The political supremacy of Great Britain throughout the nineteenth century was closely associated with London's position as the financial centre of the world. Only in Europe, where Great Britain did not aspire to political influence, were British investments insignificant, amounting to not more than 5 per cent of all British capital invested abroad. The rise of the United States to political power in the present century was largely due to their appearance in the market as a large-scale lender, first of all, to Latin America, and since 1914, to Europe. The attainment of political objectives by direct government investment occurred in such cases as the purchase by the British Government of shares in the Suez Canal Company and the Anglo-Iranian Oil Company, or the construction of the Chinese Eastern Railway with Russian Government capital. More often, governments used their power to stimulate investments by banks and private individuals in the interests of national policy. Thus the Franco-Russian alliance was cemented by some £400,000,000 of French capital lent by French investors to the Russian Government. In Germany, 'the joint-stock bank was not merely a credit organization, but a politico-economic instrument; it was an instrument of Germany's power policy'.[45] The whole policy of nineteenth-century imperialism was based on the development of the backward parts of the world through investment of European capital. Political interests were furthered by private investors enjoying, like the chartered companies of the nineteenth century, government patronage or, more commonly, diplomatic support.[46] Marx described the policy as one of replacing 'the feudal method of waging war ... by the mercantile method, cannon by capital';[47] and a new and expressive phrase was coined to describe the 'dollar diplomacy' of the United States.

> The diplomacy of the present administration [said Taft in 1912] has sought to respond to the modern ideas of commercial intercourse. This policy has been characterised as substituting dollars for bullets. It is one that appeals alike to idealistic humanitarian sentiments, to the dictates of sound policy and strategy, and to legitimate commercial aims.[48]

The frequent appearances of the American fleet in Latin American waters (like those of the British fleet elsewhere) showed, moreover, that, if dollars were a humanitarian substitute for bullets, they could and would be reinforced by bullets in case of political need.

The diminished use after 1919 of capital investment abroad as an instrument of policy was explained by the rapid falling-off in the accumulation of surplus capital throughout the world and the insolvency of many potential borrowers. But numerous familiar examples may still be cited. France strengthened her influence over Poland and the Little Entente by abundant loans and credits, public and private, to these countries. Several governments granted or guaranteed loans to Austria for the political purpose of maintaining Austria's independence; and in 1931 French financial pressure obliged Austria to abandon the project of a customs union between Austria and Germany. The rapid decline of French influence in Central Europe after 1931 was closely connected with the fact that France, since the crisis, was unable to continue her policy of financial assistance to these countries. When in December 1938 it was announced that the French Schneider-Creusot group had sold its interest in the Skoda works to a Czecho-Slovak group representing the Czecho-Slovak Government, a correspondent of *The Times* commented that 'this transaction is another indication of France's retreat from Central Europe, and puts an end to a chapter of French political expansion'.[49] After 1932, when an unofficial embargo was placed on the issue of foreign loans in the British market, it could fairly be said that Great Britain's foreign lending was subject to political supervision. The years 1938 and 1939 saw the grant to Turkey by Great Britain and Germany, and to China by the United States and Great Britain, of 'commercial' credits whose political motive was scarcely disguised.

(*b*) The struggle to control foreign markets provides a further illustration of the interaction of politics and economics; for it is often impossible to decide whether political power is being used to acquire markets for the sake of their economic value, or whether markets are being sought in order to establish and strengthen political power. The struggle for markets has been the most characteristic feature of the economic warfare of the period between the two world wars. It would be wrong to attribute exclusively to political rivalries the intensified pressure to export which manifested itself everywhere. Under the modern structure of industry, the most economical scale of production of many commodities exceeds the consumption capacity of most national markets; and to sell dear in a protected home market and cheap in a free foreign market (which is the essence of 'dumping') may be perfectly sound policy from the purely commercial standpoint. Yet the use of dumping as an instrument of policy is incontestable; and powerful countries found their 'natural' markets in areas where their political interests lay and where their political influence could be most readily asserted. The principal reason why Central and South-Eastern Europe were Germany's 'natural' markets was their accessibility to Germany's military power. German rearmament and German economic penetration of these areas proceeded simultaneously. This was, however, not a new phenomenon. An admirable example of the intertwining of political and economic power may be found in the British position in Egypt. British economic penetration in Egypt in the last two decades of the nineteenth century resulted from British military occupation, which was designed to protect British interests in the Suez Canal, which had been acquired to protect British trade routes and strategic lines of communication.

The methods used to encourage exports and capture foreign markets are too familiar to need discussion. The simplest of all is the granting of loans or credits to finance exports. Before 1914 Great Britain was so little preoccupied with the problem of markets that loans obtained in London by foreign borrowers were free of any condition as to where the proceeds should be spent. Foreign loans obtained elsewhere frequently carried the condition that the whole or part of the proceeds should be expended by the borrower in the lending country.[50] Since 1919 this condition has been almost universally applied. In Great Britain, two governmental institutions – the Colonial Development Fund and the Export Credits Guarantee Department – were engaged in financing British exports, the first to the Empire, the second to foreign countries. Before 1939, the operations of the Export Credits Guarantee Department were officially described as being of a purely commercial character. But by an act passed in 1939, the limit of the guarantees which might be given by the Department was increased, and a sum of £10,000,000 was earmarked for the guaranteeing of transactions 'in connection with which it appears to them [i.e. the Board of Trade] expedient in the national interest that guarantees should be given'.[51] In introducing this measure into the House of Commons, the President of the Board of Trade denied the suggestion that Great Britain 'had declared a trade war upon Germany', but described the measure as one of 'economic rearmament' and added that 'the economic rearmament which we are trying now to undertake is exactly like our other rearmament'.[52] In July 1939, the amount of £10,000,000 was increased to £60,000,000. Export bounties and currency manipulation are merely indirect forms of export credits.

The most characteristic modern method of acquiring markets and the political power which goes with them is, however, the reciprocal trade agreement – the return to a system of thinly disguised barter. Thus British purchases of meat and cereals in the Argentine and of bacon and butter in Denmark and the Baltic States secured markets in those countries for British coal and British manufactures. The Ottawa Agreements were a slightly more complicated variation on the same theme. In the Central European and Balkan countries Germany, by purchasing local products (mainly cereals and tobacco) for which no other lucrative outlet could be found, secured not only a market for German goods, but a sphere of political influence. One of the symptoms of the artificial character of French political influence in this region was failure to secure any substantial share in its trade. Purchasing power had become an international asset; and the fact that price was no longer the dominant factor (Germany made most of her purchases in South-Eastern Europe at rates above world prices) put the purchaser and not the producer in a position to call the tune. A new power has thus been placed in the hands of countries with a large population and a high standard of living. But it is a wasting asset which, if used to excess, tends to destroy itself.

Economic power and international morality

One concluding reflexion may round off this summary sketch of the use of the economic weapon as an instrument of political power. The substitution of the economic weapon for the military weapon – what Marx calls the replacement of cannon by capital – is a symptom not so much of superior morality as of superior strength. This can be seen from a few simple examples. Great Britain, aggrieved by the trial of the Metro-Vickers engineers in Moscow, could obtain satisfaction by imposing an embargo on Soviet imports. Italy, aggrieved by the murder of an Italian officer, could not avail herself of this economic expedient (for an Italian embargo on Greek imports would have been negligible); she could obtain satisfaction only by the brutal military method of bombarding Corfu. In 1931, Great Britain established what came to be known as a 'sterling *bloc*' by methods which were non-political and in appearance largely fortuitous. Germany, in order to establish an equivalent 'mark *bloc*' in Central and South-Eastern Europe, had to resort to methods which were frankly political and included the use and threatened use of force. British economic and financial strength enabled Great Britain to refrain from intervention in the Spanish civil war. The British Government relied on 'sterling bullets' to prevent the permanent predominance of Germany and Italy in Spain, whatever the issue of the war. As regards the Far East, the Prime Minister in the same period remarked that 'when the war is over, and the reconstruction of China begins, she cannot possibly be reconstructed without some help from this country'.[53] The growing strength of the United States in international trade and finance was one, at any rate, of the reasons which allowed the United States Government to abandon its traditional practice of landing marines in the territory of recalcitrant Latin American republics and to adopt the 'good neighbour' policy.

The point, however, has a wider application to the whole problem of 'aggression' and territorial annexation. One of the most revealing documents on this aspect of power is a despatch from the Russian Chargé d'Affaires in Peking to the Russian Government in 1910:

Should we be sufficiently powerful economically [wrote this frank diplomat], it would be simpler to direct all our efforts to the conclusion of an economic treaty. If, however, as I fear, we should by so doing only be of service to foreigners and ourselves be unable to secure any profits from what had been achieved (thus we have for instance in reality been unable to profit by the extraordinary advantages embodied in the commercial treaty of 1881), then there is, in my opinion, no reason to depart from the basis of policy we have followed hitherto, that of territorial acquisition.[54]

A recent British writer on the Far East has made a similar observation:

Free Trade, as championed by England in the nineteenth century, was the cause of the stronger in purely commercial competition. The 'sphere of influence' with its special rights was the objective of states which sought to compensate for weakness in such competition by the direct application of political power.[55]

Great Britain's unchallenged naval and economic supremacy throughout the nineteenth century enabled her to establish a commanding position in China with a minimum of military force and of economic discrimination. A relatively weak Power like Russia could only hope to achieve a comparable result by naked aggression and annexation. Japan afterwards learned the same lesson. In his well-known memorandum of January 1907, Crowe argued that Great Britain was 'the natural protector of the weaker communities', and that by her free-trade policy of an open market 'she undoubtedly strengthens her hold on the interested friendship of other nations'.[56] The argument might have been developed by adding that Great Britain, in virtue of her inherent economic strength and the free-trade policy made possible by it, was able to exercise in many countries a measure of indirect influence and control which no other Power could have achieved without interference with the political independence of the countries concerned, and that this advantage made it as natural for Great Britain, as it would have been difficult for others, to appear as a champion of the political independence of small nations. In Egypt, Great Britain has reconciled her military and economic predominance with the formal independence of the country, where a weaker Power would have had to resort to annexation to obtain a similar effect. Great Britain was able to abandon her formal authority over Iraq and to maintain her interests there, while France shrank from the same step in Syria. The economic weapon is pre-eminently the weapon of strong Powers. It is significant that a proposal made by the Soviet Government in 1931 for a pact of 'economic non-aggression' was received with the greatest hostility by the three most powerful countries of the day: Great Britain, France and the United States.

Nevertheless, it is perhaps difficult to dismiss as unfounded the common view that the use of the economic weapon is less immoral than the use of the military weapon. This may not always be true. Blockade in time of war may cause as much suffering as a series of air raids. But generally speaking, there is a sense in which dollars are more humane than bullets even if the end pursued be the same. It is less immoral to place an embargo on Soviet imports than to bombard Greeks. It cannot be reasonably doubted that a form of economic control (such as that of the United States in Central America) which preserves a measure of political independence is more acceptable to subordinate nations, and therefore less immoral, than direct political control (such as that established by Germany in 1939 in Bohemia and Moravia). The distinction is not entirely removed by pointing out that the United States, if she were economically as weak as Germany, might well have taken the same course. It is true that the poor are more likely to steal than the rich, and that this affects our moral judgment of individual cases of theft. But theft is generally recognized as *per se* immoral. This is merely an illustration of the way in which morality itself is involved in questions of power.

The moral issue will require consideration later. For the present, the most important lesson to be drawn in this field is the illusory character of the popular distinction between economic and military power. Power, which is an element of all political action, is one and indivisible. It uses military and economic weapons for the same ends. The strong will tend to prefer the minor and more 'civilized' weapon, because it will generally suffice to achieve his purposes; and as long as it will suffice, he is under no temptation to resort to the more hazardous military weapon. But economic power cannot be isolated from military power, nor military from economic. They are both integral parts of political power; and in the long run one is helpless without the other.

(c) Power over opinion

Power over opinion is the third form of power. The 'Jingoes' who sang 'We've got the ships, we've got the men, we've got the money too' had accurately diagnosed the three essential elements of political power: armaments, manpower and economic power. But manpower is not reckoned by mere counting of heads. 'The Soldan of Egypt or the Emperor of Rome', as Hume remarked, 'might drive his harmless subjects like brute beasts against their sentiments and inclinations. But he must at least have led his *mamelukes* or pretorian bands like men by their opinions.'[57] Power over opinion is therefore not less essential for political purposes than military and economic power, and has always been closely associated with them. The art of persuasion has always been a necessary part of the equipment of political leader. Rhetoric has a long and honoured record in the annals of statesmanship. But the popular view which regards propaganda as a distinctively modern weapon is, none the less, substantially correct.

Propaganda in the modern world

The most obvious reason for the increasing prominence attached to power over opinion in recent times is the broadening of the basis of politics, which has vastly increased the number of those whose opinion is politically important. Until comparatively modern times, those whose opinion it was worth while to influence were few in number, united by close ties of interest and, generally speaking, highly educated; and the means of persuasion were correspondingly limited. 'Scientific exposition', in Hitler's words, is for the intelligentsia. The modern weapon of propaganda is for the masses.[58] Christianity seems to have been the first great movement in history with a mass appeal. Appropriately enough, it was the Catholic church which first understood and developed the potentialities of power over large masses of opinion. The Catholic church in the Middle Ages was – and has, within the limits of its power, remained – an institution for diffusing certain opinions and extirpating other opinions contrary to them: it created the first censorship and the first propaganda organization. There is much point in the remark of a recent historian that the mediaeval church was the first totalitarian state.[59] The Reformation was a movement which simultaneously deprived it, in several parts of Europe, of its power over opinion, of its wealth and of the authority which the military power of the Empire had conferred on it.

The problem of power over opinion in its modern mass form has been created by developments in economic and military technique – by the substitution of mass-production industries for individual craftsmanship and of the conscript citizen army for the volunteer professional force. Contemporary politics are vitally dependent on the opinion of large masses of more or less politically conscious people, of whom the most vocal, the most influential and the most accessible to propaganda are those who live in and around great cities. The problem is one which no modern government ignores. In appearance, the attitude adopted towards it by democracies and by totalitarian states is diametrically opposed. Democracies purport to follow mass opinion; totalitarian states set a standard and enforce conformity to it. In practice, the contrast is less clear cut. Totalitarian states, in determining their policy, profess to express the will of the masses; and the profession is not wholly vain. Democracies, or the groups which control them, are not altogether innocent of the arts of moulding and directing mass opinion. Totalitarian propagandists, whether Marxist or Fascist, continually insist on the illusory character of the freedom of opinion enjoyed in democratic countries. There remains a solid substratum of difference between the attitude of democracies and totalitarian states towards mass opinion, which may prove a decisive factor in times of crisis. But both agree in recognizing its paramount importance.

The same economic and social conditions which have made mass opinion supremely important in politics have also created instruments of unparalleled range and efficiency for moulding and directing it. The oldest, and still perhaps the most powerful, of these instruments is universal popular education. The state which provides the education necessarily determines its content. No state will allow its future citizens to imbibe in its schools teaching subversive of the principles on which it is based. In democracies, the child is taught to prize the liberties of democracy; in totalitarian states, to admire the strength and discipline of totalitarianism. In both, he is taught to respect the traditions and creeds and institutions of his own country, and to think it better than any other. The influence of this early unconscious moulding is difficult to exaggerate. Marx's dictum that 'the worker has no country' has ceased to be true since the worker has passed through national schools.

But when we speak of propaganda to-day, we think mainly of those other instruments whose use popular education has made possible: the radio, the film and the popular press. The radio, the film and the press share to the fullest extent the characteristic attribute of modern industry, i.e. that mass-production, quasi-monopoly and standardization are a condition of economical and efficient working. Their management has, in the natural course of development, become concentrated in fewer and fewer hands; and this concentration facilitates and makes inevitable the centralized control of opinion. The mass-production of opinion is the corollary of the mass-production of goods. Just as the nineteenth-century conception of political freedom was rendered illusory for large masses of the population by the growth and concentration of economic power, so the nineteenth-century conception of freedom of thought is being fundamentally modified by the development of these new and extremely powerful instruments of power over opinion. The prejudice which the word propaganda still excites in many minds to-day[60] is closely parallel to the prejudice against state control of industry and trade. Opinion, like trade and industry, should according to the old liberal conception be allowed to flow in its own natural channels without artificial regulation. This conception has broken down on the hard fact that in modern conditions opinion, like trade, is not and cannot be exempt from artificial controls. The issue is no longer whether men shall be politically free to express their opinions, but whether freedom of opinion has, for large masses of people, any meaning but subjection to the influence of innumerable forms of propaganda directed by vested interests of one kind or another. In the totalitarian countries, radio, press and film are state industries absolutely controlled by governments. In democratic countries, conditions vary, but are everywhere tending in the direction of centralized control. Immense corporations are called into existence, which are too powerful and too vital to the community to remain wholly independent of the machine of government, and which themselves find it convenient to accept voluntary collaboration with the state as an alternative to formal control by it. The nationalization of opinion has proceeded everywhere *pari passu* with the nationalization of industry.

Propaganda as an instrument of policy

The organized use of power over opinion as a regular instrument of foreign policy is a modern development. Before 1914, cases occurred of the use of propaganda by governments in international relations. The press was freely used by Bismarck and other statesmen, though rather for the purpose of making pronouncements to foreign governments than as a means of influencing public opinion at large. Co-operation between the missionary and the trader, and the support of both by military force, was a familiar nineteenth-century example of unofficial association between propaganda and economic and military power in the interests of national expansion. But the field of propaganda was limited; and the only people who exploited it at all intensively were the revolutionaries. Any systematic resort to propaganda by governments would have been thought undignified and rather disreputable.

It did not take long for the belligerents of 1914–18 to realize that 'psychological war must accompany economic war and military war'.[61] It was a condition of success on the military and economic fronts that the 'morale' of one's own side should be maintained, and that of the other side sapped and destroyed. Propaganda was the instrument by which both these ends were pursued. Leaflets were dropped over the enemy lines inciting his troops to mutiny; and this procedure, like most new weapons of war, was at first denounced as being contrary to international law.[62] Moreover, the new conditions of warfare nullified, in this as in so many other respects, the distinction between combatant and civilian; and the morale of the civilian population became for the first time a military objective.

> Long-distance bombing [wrote the British Chief of Staff in January 1918] will produce its maximum moral effect only if visits are constantly repeated at short intervals so as to produce in each area bombed a sustained anxiety. It is this recurrent, as opposed to isolated spasmodic attacks, which interrupts industrial production and under-mines public confidence.[63]

The military chiefs of other belligerent countries were doubtless consider-ing the same problem in similar terms. The demoralization of the civilian population was the primary objective not only of many air raids but of the German long-range bombardment of Paris by 'big Bertha'; and the work of the bomb and the shell was reinforced, especially during the last months of the war, by an intense output of printed propaganda. Throughout the first world war the close interdependence between the three forms of power was constantly demonstrated. The success of propaganda on both sides, both at home and in neutral and enemy countries, rose and fell with the varying fortunes of the military and economic struggle. When at length the Allied blockade and Allied victories in the field crippled German resources, Allied propaganda became enormously effective and played a considerable part in the final collapse. The victory of 1918 was achieved by a skilful combination of military power, economic power and power over opinion.

Notwithstanding the general recognition of the importance of propaganda in the later stages of the war, it was still regarded by almost everyone as a weapon specifically appropriate to a period of hostilities. 'In the same way as I send shells into the enemy trenches, or as I discharge poison gas at him,' wrote the German general who was primarily responsible for despatching Lenin and his party in the sealed train to Russia, 'I, as an enemy, have the right to use propaganda against him.'[64] The abolition of ministries and departments of propaganda at the end of the war was an automatic measure of demobilization. Yet within twenty years of the armistice, in what was still formally a time of peace, many governments were conducting propaganda with an intensity unsurpassed in the war period; and new official or semi-official agencies for the influencing of opinion at home and abroad were springing up in every country. This new development was rendered possible and inevitable by the popularization of international politics and by the growing efficiency of propaganda methods. Since both these processes are likely to continue, its permanence seems assured.

The initiative in introducing propaganda as a regular instrument of international relations must be credited to the Soviet Government. The causes of this were partly accidental. The Bolsheviks, when they seized power in Russia, found themselves desperately weak in the ordinary military and economic weapons of international conflict. The principal element of strength in their position was their influence over opinion in other countries; and it was therefore natural and necessary that they should exploit this weapon to the utmost. In early days, they seriously believed in their ability to dissolve the German armies by the distribution of propaganda leaflets and by fraternization between the lines. Later, they counted on propaganda in Allied countries to paralyse Allied intervention against them in the civil war. Had not propaganda been supplemented by the creation of an effective Red Army, it might by itself have proved ineffective. But the importance of the role it played is sufficiently indicated by the fear of Bolshevik propaganda felt for many years afterwards, and not yet extinct in many European and Asiatic countries. Soviet Russia was the first modern state to establish, in the form of the Communist International, a large-scale permanent international propaganda organization.

There was, however, a profounder cause why control over opinion should have taken a foremost place in the policy of Soviet Russia. Since the end of the Middle Ages, no political organization had claimed to be the repository of universal truth or the missionary of a universal gospel. Soviet Russia was the first national unit to preach an international doctrine and to maintain an effective world-propaganda organization. So revolutionary did this innovation appear that the Communist International purported at the outset to be wholly unconnected with the power of the Soviet Government. But this separation, which may have been effective in details of administration, never extended to major issues of policy; and after the Soviet state had been consolidated under Stalin, the separation became no more than a polite fiction. This development had far more than a local significance, and gives us the clue to the whole problem of the place of what are now known as 'ideologies' in international politics. For if it be true that power over opinion cannot be dissociated from other forms of power, then it appears to follow that, if power cannot be internationalized, there can be no such thing in politics as international opinion, and international propaganda is as much a contradiction in terms as an international army. This view, paradoxical as it may appear, can be supported by extremely cogent arguments; and both it and its implications require careful examination.

National or international propaganda?

Most political ideas which have strongly influenced mankind have been based on professedly universal principles and have therefore had, at any rate in theory, an international character. The ideas of the French Revolution, free trade, communism in its original form of 1848 or in its reincarnation of 1917, Zionism, the idea of the League of Nations, are all at first sight (as they were in intention) examples of international opinion divorced from power and fostered by international propaganda. But reflexion will set limits on this first impression. How far were any of these ideas politically effective until they took on a national colour and were supported by national power? The answer is not easy. Albert Sorel has a well-known passage on the course taken by the enthusiasm of the French revolutionaries:

> They confuse ... the propagation of the new doctrines with the extension of French power, the emancipation of mankind with the greatness of the Republic, the rule of reason with that of France, the liberation of peoples with the conquest of states, the European revolution with the domination of the French Revolution over Europe.[65]

The military power of Napoleon was notoriously the most potent factor in the propagation throughout Europe of the ideas of 1789. The political influence of the idea of free trade dated from its adoption by Great Britain as the basis of British policy. The revolutionaries of 1848 failed everywhere to achieve political power; and the ideas of 1848 remained barren. Neither the First nor the Second International attained any real authority. As 1914 showed, there were national labour movements, but there was no international labour movement. The Third or Communist International enjoyed little influence until the power of the Russian state was placed behind it; and Stalin has garbled and disseminated the ideas of 1917 in much the same way as Napoleon garbled and disseminated the ideas of 1789. Trotskyism, unsupported by the power of any state, remains without influence. Zionism, politically impotent so long as it relied solely on international propaganda, is effective in so far as it can count on the political backing of Great Powers. Propaganda is ineffective as a political force until it acquires a national home and becomes linked with military and economic power.

The fate of the League of Nations and of propaganda on its behalf is perhaps the best modern illustration of this tendency. As has been shown, men like Woodrow Wilson and Lord Cecil conceived the League of Nations as an expression of 'the organized opinion of mankind' controlling the military and economic power of governments. International public opinion was the supreme instrument of power ('by far the strongest weapon we have'); and this opinion was to be created by international propaganda which took no heed of frontiers.* Throughout the nineteen-twenties, this fallacy of the power of international opinion was being gradually exposed. That it survived at all was due to the persistent use by League enthusiasts of slogans like peace and disarmament which were capable of a universal appeal precisely because they meant different, and indeed contradictory, things to different people. Every country wanted to achieve the aims of its policy without war, and therefore stood for peace. Every country wanted disarmament of other countries or disarmament in those weapons which it did not regard as vital to itself. After the collapse of the Disarmament Conference, it became apparent to all that the League of Nations could be effective only in so far as it was an instrument of the national policy of its most powerful members. Opinion in favour of the League ceased altogether to be international, and was confined to those countries where the League was felt to be serving ends of national policy. In Great Britain the League of Nations became for the first time popular with what might be called the nationalist wing of the Conservative Party.

The fallacy of belief in the efficacy of an international public opinion divorced from national power may be further illustrated by developments elsewhere. The group of movements conveniently classified under the rubric of Fascism was based on certain professedly universal principles such as the rejection of democracy and class warfare, the insistence on leadership, and so forth. In its early days, Fascism was authoritatively described as 'not an article for export', and was for many years so treated by the countries which adhered to it. At a later date this limitation was explicitly disclaimed,[66] and Fascism became the theme of a vigorous

*See pp. 32–5.

international propaganda in many parts of the world. It would, however, be a superficial diagnosis to pretend that, while the League of Nations and the Communist International began as instruments of international opinion and ended as instruments of national policy, Fascism began as an instrument of national policy and ended as an instrument of international opinion. In both cases, the international phase was an illusion (which does not mean that many people may not sincerely believe in it). International propaganda for Fascism was an instrument of the national policy of certain states, and grew with the growth of the military and economic power of those states. But the *reductio ad absurdum* of international ideological propaganda as a cloak for national policy came with the adoption of negative slogans designed to unite in a political alliance those who shared no positive ideology in common. Thus the Anti-Comintern Pact did not prevent Germany from coming to an agreement with the principal Communist Power when the needs of national policy seemed to require it; and the 'anti-Fascism' of the democratic nations did not deter them from seeking the alliance of countries whose forms of government were indistinguishable from Fascism. These slogans had no meaning or substance apart from the national policies of the countries by which they were used. Power over opinion cannot be dissociated from military and economic power.

International agreements regarding propaganda

Propaganda is now so well recognized as a national political weapon that stipulations regarding its use are fairly common in international agreements. Such stipulations were, appropriately enough, first introduced into agreements made with the Soviet Government for the purpose of limiting the activities of the Communist International. But this could still be thought of as an exceptional case. Outside Soviet Russia, the first recorded agreement to abstain from hostile propaganda seems to have been one concluded between the German and Polish Broadcasting Companies, which undertook to assure that 'the matter broadcast does not in any way offend the national sentiment of listeners who are nationals of the other contracting party'.[67] Propaganda was first raised to the dignity of a universal issue when the Polish Government made proposals to the Disarmament Conference for a convention on 'moral disarmament'. To limit the propaganda weapon by a general convention proved as hopeless a task as to limit the military weapon.[68] But bilateral agreements for terminating hostile propaganda were concluded between Germany and Poland in 1934 and between Germany and Austria in 1936;[69] and in the Anglo-Italian Agreement of 16 April 1938, the two countries 'placed on record their agreement that any attempt by either of them to employ the methods of publicity and propaganda at its disposal in order to injure the interests of the other would be inconsistent with the good relations which it is the object of the present agreement to establish'.

Such agreements create an obvious difficulty for democracies, which purport not to limit the free expression and publication of opinions about international affairs, and cannot therefore formally undertake to prevent propaganda on their territory against any country; and this embarrassment is reflected in the contorted phraseology of the Anglo-Italian Agreement. The fact is, however, that in the sphere of opinion, as in the economic sphere, the nineteenth-century principles of *laissez-faire* no longer hold good, even for democracies. Just as democratic governments have been compelled to control and organize economic life in their territories in order to compete with totalitarian states, so they find themselves at a disadvantage in dealing with these states if they are not in a position to control and organize opinion. Recognition of this fact grew rapidly even in Great Britain. In questions affecting international relations, a discreet influence, amounting in times of crisis to direct though unofficial censorship, was exercised even before the outbreak of the second world war over broadcasting, films and press; and though the use of this influence was frequently criticized in particular cases, it became clear that some such measures of restraint would be applied in similar circumstances by whatever government happened to be in power.[70] Simultaneously, there was a rapid extension of propaganda designed to familiarize foreign opinion with the British point of view. Since 1935, a body called the British Council has exercised the function of 'making the life and thought of the British peoples more widely known abroad'. In 1938, the British Broadcasting Corporation began the regular broadcasting of news bulletins in various foreign languages. In June 1939, the Prime Minister announced the creation of a new Foreign Publicity Department of the Foreign Office, which served as a nucleus for the Ministry of Information set up immediately on the outbreak of war.

Truth and morality in propaganda

We have hitherto discussed power over opinion in precisely the same terms as military and economic power; and the close connexion between these different forms of power is so vital, and has been so much neglected in theoretical discussion, that this seems the most fruitful approach to the problem at the present time. Some people might indeed argue that this is the only correct approach. For in the first place, opinion is conditioned by status and interest; and secondly, as we have seen in a previous chapter, a ruling class or nation, or dominant group of nations, not only evolves opinions favourable to the maintenance of its privileged position, but can, in virtue of its military and economic superiority, easily impose these opinions on others. The victory of the democratic countries in 1918 created an almost universal opinion that democracy was the best form of government. In the nineteen-thirties, opinion in many parts of the world on the merits of Fascism as a form of government may be said, without much exaggeration, to have varied *pari passu* with the military and economic power of Germany and Italy in relation to the other Great Powers. These propositions could be supported by innumerable examples. If they were absolutely true, then power over opinion would in fact be indistinguishable in character from military and economic power, and there would be nothing which, given sufficient power and technical skill, men could not be made to believe. That this is the case has indeed sometimes been suggested. 'By clever, persistent propaganda', said Hitler, 'even heaven can be represented to a people as hell, and the most wretched life as paradise';[71] and American advertising specialists are alleged to hold that 'only cost limits the delivery of public opinion in any direction on any topic'.[72] But these are the pardonable exaggerations of expert practitioners. As we shall see, even Hitler did not really believe in the unlimited power of propaganda to manufacture opinion. Here as elsewhere, the extreme realist position becomes untenable. When we set power over opinion side by side with military and economic power, we have none the less to remember that we are dealing no longer with purely material factors, but with the thoughts and feelings of human beings.

Absolute power over opinion is limited in two ways. In the first place, it is limited by the necessity of some measure of conformity with fact. There are objective facts which are not totally irrelevant to the formation of opinion. Good advertising may persuade the public that a face cream made of inferior materials is the best. But the most expert advertiser could not sell a face cream made of vitriol. Hitler condemned the futility of German propaganda in the first world war which depicted the enemy as ridiculous and contemptible. The propaganda was unsuccessful simply because it was, as the German soldier in the trenches discovered, untrue. This danger that 'truth will out', especially in an age of competitive propaganda, is a serious limitation on power over opinion. Education, which is one of the strongest instruments of this power, tends at the same time to promote a spirit of independent enquiry which is also one of the strongest antidotes against it. In so far as it strains and interprets facts for a specific purpose, propaganda always contains within itself this potentially self-defeating element.

Secondly, power over opinion is limited – and perhaps even more effectively – by the inherent utopianism of human nature. Propaganda, harnessed to military and economic power, always tends to reach a point where it defeats its own end by inciting the mind to revolt against that power. It is a basic fact about human nature that human beings do in the long run reject the doctrine that might makes right. Oppression some-times has the effect of strengthening the will, and sharpening the intelligence, of its victims, so that it is not universally or absolutely true that a privileged group can control opinion at the expense of the unprivileged. As Hitler himself wrote, 'every persecution which lacks a spiritual basis' has to reckon with a 'feeling of opposition to the attempt to crush an idea by brute force'.[73] And this vital fact gives us another clue to the truth that politics cannot be defined solely in terms of power. Power over opinion, which is a necessary part of all power, can never be absolute. International politics are always power politics; for it is impossible to eliminate power from them. But that is only part of the story. The fact that national propaganda everywhere so eagerly cloaks itself in ideologies of a professedly international character proves the existence of an interna-tional stock of common ideas, however limited and however weakly held, to which appeal can be made, and of a belief that these common ideas stand somehow in the scale of values above national interests. This stock of common ideas is what we mean by international morality.

Notes

1. Lenin, *Selected Works* (Engl. transl.), vii. p. 295.
2. Marx and Engels, *Works* (Russian ed.), vii. p. 212.
3. Even Lord Baldwin committed himself in 1925 to the dangerous half-truth that 'democracy is government by discussion, by talk' (*On England*, p. 95). In a recent letter to *The Times*, Mr Frederic Harrison remarks of the British Commonwealth of Nations that it 'is not founded on conquest and held together by force of arms. It has been acquired not by the force of our navy and our army but by force of character, and knit together by ties of sympathy, of a common interest, a common language and a common history' (*The Times*, 30 June 1938). This, too, is a dangerous half-truth, which burkes the other and equally important half of the truth, i.e. that the British Commonwealth is held together by the immense military and economic power of Great Britain and would at once dissolve if that power were lost.
4. Report to the Sixteenth Congress of the Russian Communist Party reprinted in *L'Union Soviétique et la Cause de la Paix*, p. 25; *The Times*, 26 June 1939. The italics have been inserted in both cases.
5. *Intimate Papers of Colonel House*, ed. C. Seymour, iv. p. 24.
6. Miller, *The Drafting of the Covenant*, ii. p. 61. The result of the subsequent enlargements of the Council has already been mentioned (p. 40 note 12).
7. *The Foreign Policy of the Powers* (1935: reprinted from *Foreign Affairs*), pp. 86–7.

8. *League of Nations: Sixteenth Assembly*, Part II. p. 49.
9. In 1926, when Palestine was discussed by the Mandates Commission, M. Rappard 'thought that the Mandatory would incur grave responsibility if it found itself one day faced with the impossibility of preventing a pogrom owing to insufficient troops. Its responsibility, indeed, would be shared by the Mandates Commission, if that Commission had not pointed out this danger' (*Permanent Mandates Commission, Minutes of Ninth Session*, p. 184). The responsibility of the Commission was thus limited to 'pointing out'.
10. Lugard, *The Dual Mandate in Tropical Africa*, p. 53.
11. B. Russell, *Power*, p. 11. I owe to this book, which is an able and stimulating analysis of power as 'the fundamental concept in social science', the tripartite classification of power adopted above.
12. Lenin, *Collected Works* (Engl. transl.), xviii. p. 97; Theses of the Sixth Congress of Comintern quoted in Taracouzio, *The Soviet Union and International Law*, p. 436.
13. Hitler, *Mein Kampf*, p. 749.
14. R. G. Hawtrey, *Economic Aspects of Sovereignty*, p. 107.
15. It is perhaps necessary to recall the part played in British politics in 1914 by the threat of the Conservative Party to support revolutionary action in Ulster.
16. R. G. Hawtrey, *Economic Aspects of Sovereignty*, p. 105.
17. *League of Nations: Official Journal*, May 1924, p. 578.
18. R. Niebuhr, *Moral Man and Immoral Society*, p. 42.
19. Machiavelli, *Discorsi*, I. i. ch. v.
20. Hobbes, *Leviathan*, ch. xi.
21. *British and Foreign State Papers*, ed. Hertslet, xc. p. 811.
22. The distinction between the two systems is implicit in Saint-Simon's prediction that the 'industrial régime' will succeed the 'military régime', and 'administration' replace 'government', better known in the form given to it by Engels that the 'administration of things' will replace the 'government of men' (quotations in Halévy, *Ère des Tyrannies*, p. 224).
23. Angell, *The Great Illusion*, ch. ii.
24. B. Bosanquet, *Social and International Ideals*, pp. 234–5.
25. Hitler, *Mein Kampf*, p. 158.
26. Engels, *Anti-Dühring* (Engl. transl.), p. 195.
27. Planned economy has been developed not only by international frictions, but by social frictions within the state. It can therefore be logically regarded both as a nationalist policy ('economic nationalism') and as a socialist policy. The second aspect was irrelevant to my present argument, and has therefore been passed over in the text. According to Bruck (*Social and Economic History of Germany*, p. 157), the term *Planwirtschaft* was invented in Germany during the first world war. But the phrase *der staatliche Wirtschaftsplan* occurs in a composite *Grundriss der Sozialökonomik* (i. 424), published at Tübingen just before the war, in the general sense of 'state economic policy'.
28. Moeller van den Bruck, *Germany's Third Empire*, p. 50. The idea is a commonplace of National Socialist and Fascist writers.

29. In Germany, 'political economy' was at first translated *Nationalökonomie*, which was tentatively replaced in the present century by *Sozialökonomie*.
30. 'Economic conflicts and divergence of economic interest are perhaps the most serious and most permanent of all the dangers which are likely to threaten the peace of the world' (*League of Nations*: C.E.I. 44, p. 7).
31. 'I have deliberately debarred myself from touching on the strictly political aspects. ... It is, however, impossible to ignore the fact that we are working in their shadow' (*Report ... on the Possibility of Obtaining a General Reduction of the Obstacles to International Trade*, Cmd 5648).
32. All these passages are quoted from the Report of the Palestine Royal Commission of 1937, Cmd 5479, pp. 298–300.
33. This interpretation is confirmed by the report of the Phillimore Committee, on whose proposals the text of Article 16 was based. The Committee 'considered financial and economic sanctions as being simply the contribution to the work of preventing aggression which might properly be made by countries which were not in a position to furnish actual military aid' (*International Sanctions: Report by a Group of Members of the Royal Institute of International Affairs*, p. 115, where the relevant texts are examined).
34. House of Commons, 18 May 1934; *Official Report*, col. 2139.
35. It is not, of course, suggested that the military weapon must always be used. The British Grand Fleet was little used in the first world war. But it would be rash to assume that the result would have been much the same if the British Government had not been prepared to use it. What paralysed sanctions in 1935–36 was the common knowledge that the League Powers were not prepared to use the military weapon.
36. It is worth noting that Stresemann was fully alive to this point when Germany entered the League of Nations. When the Secretary-General argued that Germany, if she contracted out of military sanctions, could still participate in economic sanctions, Stresemann replied: 'We cannot do that either; if we take part in an economic boycott of a powerful neighbour, a declaration of war against Germany might be the consequence, since the exclusion of another country from intercourse with a nation of sixty million citizens would be a hostile act' (*Stresemann's Diaries and Papers* (Engl. transl.), ii. p. 69).
37. F. L. Schuman, *International Politics*, p. 356.
38. Zimmern, *Quo Vadimus?*, p. 41.
39. *Works of Alexander Hamilton*, iv. pp. 69 *sqq*.
40. List, *The National System of Political Economy* (Engl. transl.), p. 425.
41. *Supply of Food and Raw Materials in Time of War*, Cmd 2644.
42. The resolutions are printed in *History of the Peace Conference*, ed. Temperley, v. pp. 368–9.
43. W. Y. Elliott in *Political Quarterly*, April–June 1938, p. 181.
44. G. D. H. Cole in *Political Quarterly*, January–March 1939, p. 65.
45. W. F. Bruck, *Social and Economic History of Germany*, p. 80.
46. The whole subject is thoroughly investigated, and innumerable examples are cited, in Eugene Staley, *War and the Private Investor*. Mr Staley's main

conclusion is that official policy has rarely been influenced in an important degree by private investment, but that private investment has again and again been officially directed and encouraged as an instrument of policy.

47. Marx, *Gesammelte Schriften*, i. p. 84.
48. Annual Presidential Message to Congress, 3 December 1912.
49. *The Times*, 29 December 1938.
50. Examples from France and Austria are quoted by C. K. Hobson, *The Export of Capital* (1914), p. 16. Russia and Belgium were also lending countries which commonly imposed this condition.
51. It may be significant that in 1938 an official of the Foreign Office was transferred to the staff of the Export Credits Guarantee Department.
52. House of Commons, 15 December 1938: *Official Report*, col. 2319.
53. House of Commons, 1 November 1938, reprinted in N. Chamberlain, *The Struggle for Peace*, p. 340.
54. B. de Siebert, *Entente Diplomacy of the World War*, p. 20.
55. G. F. Hudson, *The Far East in World Politics*, p. 54.
56. *British Documents on the Origins of the War*, ed. Gooch and Temperley, iii. p. 403.
57. *The Philosophical Works of David Hume*, iv. p. 31.
58. Hitler, *Mein Kampf*, p. 196.
59. G. G. Coulton, *Mediaeval Panorama*, p. 458 *et al.*
60. 'I wish', said the Home Secretary in the House of Commons on 28 July 1939, 'there had been no necessity for any Government publicity anywhere in the world. I still look forward to living long enough to see an end of this objectionable relic of the years of the war' (*Official Report*, col. 1834).
61. H. D. Lasswell in the Foreword to G. G. Bruntz, *Allied Propaganda and the Collapse of the German Empire*. This book is the most comprehensive available account of its subject.
62. In 1917, two British airmen captured by the Germans were sentenced to ten years' hard labour for dropping such leaflets in contravention of the laws of war. The sentences were remitted on a British threat of reprisals. The practice was explicitly sanctioned in The Hague rules of 1923 for the conduct of aerial warfare (Bruntz, *op. cit.*, pp. 142–4).
63. *The War in the Air* (British Official History of the War), by H. A. Jones vi. Appendix VI, p. 26.
64. Hoffmann, *War Diaries* (Engl. transl.), ii. p. 176.
65. A Sorel, *L'Europe et la Révolution Française*, pp. 541–2.
66. Mussolini, *Scritti e Discorsi*, vi. 151; vii. 230.
67. *League of Nations*, C. 602, M. 240, 1931, ix, p. 4.
68. An international convention under which the parties undertook to prevent the broadcasting from their territories of 'incitements to war', or in general hostile propaganda, against other contracting parties, was signed at Geneva by most of the surviving members of the League in September 1936 (*League of Nations*, C. 399 (1), M. 252 (1), 1936, xii).
69. In both cases the agreement about propaganda did not figure in an officially published text, but its existence was disclosed in communiqués. The

communiqué of the Austrian Foreign Office on the German–Austrian Agreement of 11 July 1936, announced that 'both countries are to refrain from all aggressive uses of the wireless, films, news services and the theatre' (*Documents on International Affairs, 1936*, p. 324).

70. A revealing debate on the press, initiated by the Liberal Opposition, took place in the House of Commons on 7 December 1938. While Liberal speakers argued for the freedom of the press on familiar nineteenth-century lines, the spokesman of the Labour Opposition declared that the freedom of the press was already illusory, and wanted to 'make every newspaper in the country responsible for every item of news it prints and answerable to this House or some public authority' (*Official Report*, col. 1293).

71. Hitler, *Mein Kampf*, p. 302.

72. J. Truslow Adams, *The Epic of America*, p. 360.

73. Hitler, *Mein Kampf*, p. 187.

CHAPTER NINE

Morality in International Politics

The place of morality in international politics is the most obscure and difficult problem in the whole range of international studies. Two reasons for its obscurity, one general and one particular, may be suggested.

In the first place, most discussions about morality are obscured by the fact that the term is commonly used to connote at least three different things:

(i) The moral code of the philosopher, which is the kind of morality most rarely practised but most frequently discussed.

(ii) The moral code of the ordinary man, which is sometimes practised but rarely discussed (for the ordinary man seldom examines the moral assumptions which underlie his actions and his judgements and, if he does, is peculiarly liable to self-deception).

(iii) The moral behaviour of the ordinary man, which will stand in fairly close relation to (ii), but in hardly any relation at all to (i).

It may be observed that the relationship between (ii) and (iii) is mutual. Not only is the behaviour of the ordinary man influenced by his moral code, but his moral code is influenced by the way in which ordinary men, including himself, behave. This is particularly true of the ordinary man's view of political morality, which tends, more than personal morality, to be a codification of existing practice, and in which the expectation of reciprocity always plays an important part.

The monopoly of international studies between the two wars by the utopian school resulted in a concentration of interest on discussions of the question what international morality ought ideally to be. There was little discussion of the moral behaviour of states except to pass hasty and sweeping condemnation on it in the light of this ideal morality. There was no discussion at all of the assumptions of the ordinary man about international morality. This was particularly unfortunate at a period in which the popularization of politics for the first time made the assumptions of the ordinary man a matter of primary importance; and the ever-widening rift between the international utopia and international reality might have been described in terms of this divergence between the theory of the philosopher and practice based on the unexpressed and often unconscious assumptions of the ordinary man. Moreover, utopia met its usual fate in becoming, unknown to itself, the tool of vested interests. International morality, as expounded by most contemporary Anglo-Saxon writers, became little more than a convenient weapon for belabouring those who assailed the *status quo*. Here as elsewhere, the student of international politics cannot wholly divest himself of utopianism. But he will be well advised to keep his feet on the ground and rigorously maintain contact between his ambitions for the future and the realities of the present. Nor should this be too difficult. The anthropologist who investigates the moral codes and behaviour of a cannibal tribe probably starts from the presupposition that cannibalism is undesirable, and is conscious of the desire that it should be abolished. But he may well be sceptical of the value of denunciations of cannibalism, and will in any case not mistake such denunciations for a scientific study of the subject. The same clarity of thought has not always distinguished students of international morality, who have generally preferred the role of the missionary to that of the scientist.

The second obscurity is peculiar to the international field. Strange as it may appear, writers on international morality are not agreed among themselves – and are not always clear in their own minds – whether the morality which they wish to discuss is the morality of states or the morality of individuals. This point is so vital to the whole discussion that it must be cleared up on the threshold of our enquiry.

The nature of international morality

The period of absolute personal rule in which the modern state first began to take shape was not much troubled by distinction between personal and state morality. The personal responsibility of the prince for acts of state could be assumed without any undue straining of the facts. Charles I may have been a good father and a bad king. But in both capacities, his acts could be treated as those of an individual.[1] When, however, the growing complication of the state machine and the development of constitutional government made the personal responsibility of the monarch a transparent travesty, the personality (which seemed a necessary condition of moral responsibility) was transferred from the monarch to the state. Leviathan, as Hobbes said, is an 'Artificial Man'. This was an important step forward. It was the personification of the state which made possible the creation of international law on the basis of natural law. States could be assumed to have duties to one another only in virtue of the fiction which treated them as if they were persons. But the personification of the state was a convenient way of conferring on it not merely duties, but rights; and with the growth of state power in the nineteenth and twentieth centuries state rights became more conspicuous than state duties. Thus the personification of the state, which began as a liberal and progressive device, came to be associated with the assertion of unlimited rights of the state over the individual and is now commonly denounced as reactionary and authoritarian. Modern utopian thinkers reject it with fervour,[2] and are consequently led to deny that morality can be attributed to the state. International morality must, on this view, be the morality of individuals.

The controversy about the attribution of personality to the state is not only misleading, but meaningless. To deny personality to the state is just as absurd as to assert it. The personality of the state is not a fact whose truth or falsehood is a matter for argument. It is what international lawyers have called 'the postulated nature' of the state.[3] It is a necessary fiction or hypothesis – an indispensable tool devised by the human mind for dealing with the structure of a developed society.[4] It is theoretically possible to imagine a primitive political order in which individuals are individuals and nothing more, just as it is possible to imagine an economic order in which all producers and traders are individuals. But just as economic development necessitated resort to the fiction of corporate responsibility in such forms as that of the joint-stock company, so political development necessitated the fiction of the corporate responsibility of the state. Nor are the rights and obligations of these fictitious entities regarded as purely legal. A bank is praised for generosity to its employees, an armaments firm is attacked for unpatriotic conduct, and railways have 'obligations to the public' and demand a 'square deal' – all issues implying the relevance, not merely of legal, but of moral standards. The fiction of the group-person, having moral rights and obligations and consequently capable of moral behaviour, is an indispensable instrument of modern society; and the most indispensable of these fictitious group-persons is the state. In particular, it does not seem possible to discuss international politics in other terms. 'Relations between Englishmen and Italians' is not a synonym for 'relations between Great Britain and Italy'. It is a curious and significant paradox that those utopian writers on international affairs who most vigorously denounce the personification of the state as absurd and sinister none the less persistently allocate moral praise and blame (generally the latter) to those imaginary entities, 'Great Britain', 'France' and 'Italy', whose existence they deny.

Continuity is another element in society which makes the fiction of the group-person indispensable. The keenest objectors to the personification of the state will have no qualms about celebrating the 150th anniversary of *The Times* or the 38th victory of 'Cambridge' in the boat race, and will confidently expect 'the London County Council' to repay, fifty years hence, money which 'it' borrows and spends to-day. Personification is the category of thought which expresses the continuity of institutions; and of all institutions the state is the one whose continuity it is most essential to express. The question whether the Belgian Guarantee Treaty of 1839 imposed an obligation on Great Britain to assist Belgium in 1914 raised both legal and moral issues. But it cannot be intelligently discussed except by assuming that the obligation rested neither personally on Palmerston who signed the treaty of 1839, nor personally on Asquith and Grey who had to decide the issue in 1914, neither on all individual Englishmen alive in 1839, nor on all individual Englishmen alive in 1914, but on that fictitious group-person 'Great Britain', which was regarded as capable of moral or immoral behaviour in honouring or dishonouring an obligation.[5] In short, international morality is the morality of states. The hypothesis of state personality and state responsibility is neither true nor false, because it does not purport to be a fact, but a category of thought necessary to clear thinking about international relations. It is true that another moral issue was also raised in 1914 – the obligation of individual Englishmen. But this was an obligation to 'Great Britain', arising out of the obligation of 'Great Britain' to 'Belgium'. The two obligations were distinct; and confused thinking is the inevitable penalty of failure to distinguish between them.

Curiously enough, this distinction seems to present more difficulty to the philosopher than to the ordinary man, who readily distinguishes between the obligation of the individual to the state, and the obligation of the state to another state. In 1935, the Opposition in the House of Commons denounced the Hoare–Laval Plan as 'a terrible crime'. But it did not denounce Sir S. Hoare as a criminal or regard him as such; it found him guilty only of an error of judgement. In 1938, some Englishmen felt 'ashamed' of the Munich Agreement. They were not 'ashamed' of themselves; for they would have done anything in their power to prevent it. They were not 'ashamed' of Mr Chamberlain; for most of them admitted that he had acted honestly, though mistakenly, and one does not feel 'ashamed' of anyone who commits an honest mistake. They were 'ashamed' of 'Great Britain', whose reputation had, in their view, been lowered by a cowardly and unworthy act. In both these cases, the same act which (in the view of the critics) represented an intellectual failure on the part of the individual represented a moral failure on the part of 'Great Britain'. The *mot* became current that the British loan of £10,000,000 to Czecho-Slovakia was 'conscience money'. The essence of 'conscience money' is that it is paid by a moral delinquent; and the moral delinquent who paid the £10,000,000 was not Mr Chamberlain, and not those individual Englishmen who had applauded the Munich Agreement, but

'Great Britain'. The obligation of the state cannot be identified with the obligation of any individual or individuals; and it is the obligations of states which are the subject of international morality.

Two objections are commonly raised to this view.

The first is that the personification of the state encourages the exaltation of the state at the expense of the individual. This objection, though it accounts for the disfavour into which the personification of the state has fallen among liberal thinkers, is trivial. The personification of the state is a tool; and to decry it on the ground of the use to which it is sometimes put is no more intelligent than to abuse a tool for killing a man. The tool can equally well be put to liberal uses through emphasis on the duty of the state both to the individual and to other states. Nor can democracy altogether dispense with personification as a means to emphasize the duty of the individual. The most sophisticated of us would probably shrink from paying taxes to a group of individual fellow-citizens, though we pay them with comparative alacrity to a personified state. The same applies with greater force to graver sacrifices. 'You would never have got young men to sacrifice themselves for so unlucky a country as Ireland,' said Parnell, 'only that they pictured her as a woman.'[6] 'Who dies if England live?' is not adequately paraphrased by 'Who dies if other Englishmen live?' Moreover, it is difficult to see how orderly international relations can be conducted at all unless Englishmen, Frenchmen and Germans believe (however absurd the belief may be) that 'Great Britain', 'France' and 'Germany' have moral duties to one another and a reputation to be enhanced by performing those duties. The spirit of international relations seems more likely to be improved by stimulating this belief than by decrying it. In any case, it is clear that human society will have to undergo a material change before it discovers some other equally convenient fiction to replace the personification of the political unit.

The second objection is more serious. If international morality is the morality of fictitious entities, is it not itself fictitious and unreal? We can at once accept the view that moral behaviour can only proceed from individuals. To deny that 'relations between Great Britain and Italy' means the same as 'relations between Englishmen and Italians' is not to deny that 'relations between Great Britain and Italy' depend on the actions of individual Englishmen and Italians. The moral behaviour of the state is a hypothesis; but we need not regard as 'unreal' a hypothesis which is accepted in certain contexts as a guide to individual behaviour and does in fact influence that behaviour. So long as statesmen, and others who influence the conduct of international affairs, agree in thinking that the state has duties, and allow this view to guide their action, the hypothesis remains effective. The acts with which international morality is concerned are performed by individuals not on their own behalf, but on behalf of those fictitious group persons 'Great Britain' and 'Italy', and the morality in question is the morality attributed to those 'persons'. Any useful examination of international morality must start from recognition of this fact.

Theories of international morality

Before we consider the moral assumptions which underlie current thinking about international affairs, we must take some account of current theories of international morality. For though it is the assumptions of the ordinary man, not the assumptions of the philosopher, which determine the accepted moral code and govern moral behaviour, the theories of philosophers also exercise an influence on the thought (and, less frequently, on the action) of the ordinary man, and cannot be left altogether out of the picture. Theories of international morality tend to fall into two categories. Realists – and, as we have seen, some who are not realists – hold that relations between states are governed solely by power and that morality plays no part in them. The opposite theory, propounded by most utopian writers, is that the same code of morality is applicable to individuals and to states.

The realist view that no ethical standards are applicable to relations between states can be traced from Machiavelli through Spinoza and Hobbes to Hegel, in whom it found its most finished and thorough-going expression. For Hegel, states are completely and morally self-sufficient entities; and relations between them express only the concordance or conflict of independent wills not united by any mutual obligation. The converse view that the same standard is applicable to individuals and to states was implicit in the original conception of the personification of the state and has found frequent expression not only in the writings of philosophers, but in the utterances of statesmen of utopian inclinations. 'The moral law was not written for men alone in their individual character,' said Bright in a speech on foreign policy in 1858, '... it was written as well for nations.'[7] 'We are at the beginning of an age', said Woodrow Wilson in his address to Congress on the declaration of war in 1917, 'in which it will be insisted that the same standards of conduct and of responsibility for wrong shall be observed among nations and their governments that are observed among the individual citizens of civilized states.'[8] And when in July 1918 the faithful House tried his hand at the first draft of a League of Nations, Article I ran as follows:

> The same standards of honour and ethics shall prevail internationally and in affairs of nations as in other matters. The agreement or promise of a power shall be inviolate.[9]

No corresponding pronouncement was included in the Covenant. But Dr Benes at one of the early Assemblies remarked that the League was '*ipso facto* an attempt to introduce into international relationships the principles and methods employed ... in the mutual relations of private individuals'.[10] In his famous Chicago speech of 5 October 1937, President Roosevelt declared that 'national morality is as vital as private morality'.[11] But he did not specifically identify them.

Neither the realist view that no moral obligations are binding on states, nor the utopian view that states are subject to the same moral obligations as individuals, corresponds to the assumptions of the ordinary man about international morality. Our task is now to examine these assumptions.

Ordinary assumptions about international morality

It is noteworthy that the attempt to deny the relevance of ethical standards to international relations has been made almost exclusively by the philosopher, not by the statesman or the man in the street. Some recognition of an obligation to our fellow-men as such seems implicit in our conception of civilization; and the idea of certain obligations automatically incumbent on civilized men has given birth to the idea of similar (though not necessarily identical) obligations incumbent on civilized nations. A state which does not conform to certain standards of behaviour towards its own citizens and, more particularly, towards foreigners will be branded as 'uncivilized'. Even Hitler in one of his speeches declined to conclude a pact with Lithuania 'because we cannot enter into political treaties with a state which disregards the most primitive laws of human society';[12] and he frequently alleged the immorality of Bolshevism as a reason for excluding Soviet Russia from the family of nations. All agree that there is an international moral code binding on states. One of the most important and most clearly recognized items in this code is the obligation not to inflict *unnecessary* death or suffering on other human beings, i.e. death or suffering not necessary for the attainment of some higher purpose which is held, rightly or wrongly, to justify a derogation from the general obligation. This is the foundation of most of the rules of war, the earliest and most developed chapter of international law; and these rules were generally observed in so far as they did not impede the effective conduct of military operations.[13] A similar humanitarian motive inspired international conventions for the protection of the 'backward races' or of national minorities, and for the relief of refugees.

The obligations so far mentioned have been obligations of the state to individuals. But the obligation of state to state is also clearly recognized. The number of synonyms current in international practice for what used to be called 'the comity of nations'[14] shows the persistence of the belief that states are members of a comity and have obligations as such. A new state on becoming, in virtue of recognition by other Powers, a member of the international community, is assumed to regard itself as automatically bound, without any express stipulation, by the accepted rules of international law and canons of international morality. As we have seen, the concept of internationalism was so freely used between the two wars for the purpose of justifying the ascendancy of the satisfied Powers that it fell into some disrepute with the dissatisfied Powers. But this natural reaction was not a denial of the existence of an international community so much as a protest against exclusion from the privileges of membership. The result of the Versailles Treaty, wrote Dr Goebbels, was 'to expel Germany from the comity of powerful political countries', and the function of National Socialism was to 'unite the people and once more lead it back to its rightful place in the comity of nations'.[15] During Hitler's visit to Rome in May 1938, Mussolini declared that the common aim of Italy and Germany was 'to seek between them and with others a regime of international comity which may restore equally for all more effective guarantees of justice, security and peace'.[16] Constant appeals were made by both these Powers to the injustice of the conditions imposed on them in the past and the justice of demands now made by them; and many people in these countries were beyond doubt sincerely and passionately concerned to justify their policy in the light of universal standards of international morality.

In particular, the theory that, since states have no moral obligations towards one another, treaties have no binding force, is not held even by those statesmen who exhibit least taste for international co-operation. Every state concludes treaties in the expectation that they will be observed; and states which violate treaties either deny that they have done so, or else defend the violation by argument designed to show that it was legally or morally justified. The Soviet Government in the first years of its existence openly violated not only treaties signed by previous Russian governments, but the treaty which it had itself signed at Brest-Litovsk, and propounded a philosophy which seemed to deny international obligation and international morality. But it simultaneously concluded, and offered to conclude, other treaties with the manifest intention of observing them and expecting others to observe them. The German Government accompanied its violation of the Locarno Treaty in 1936 with an offer to enter into a fresh treaty. In neither case is it necessary to doubt the sincerity of the government concerned. Violation of treaties, even when frequently practised, is felt to be something exceptional requiring special justification. The general sense of obligation remains.

The view that the same ethical standard is applicable to the behaviour of states as to that of individuals is, however, just as far from current belief as the view that no standard at all applies to states. The fact is that most people, while believing that states ought to act morally, do not expect of them the same kind of moral behaviour which they expect of themselves and one another.

Many utopian thinkers have been so puzzled by this phenomenon that they have refused to recognize it. Others have sincerely confessed their bewilderment. 'Men's morals are paralysed when it comes to international conduct,' observes Professor Dewey;[17] and Professor Zimmern detects a 'rooted prejudice against law and order in the international domain'.[18] The discrepancy is less surprising than it appears at first sight. Casuists have long been familiar with the problem of incompatibilities between personal, professional and commercial morality. International morality is another category with standards which are in part peculiar to itself. Some of the problems of state morality are common to the whole field of the morality of group persons. Others are peculiar to the state in virtue of its position as the supreme holder of political power. The analogy between the state and other group persons is therefore useful, but not decisive.

Differences between individual and state morality

We may now turn to the principal reasons why states are not ordinarily expected to observe the same standards of morality as individuals.

(1) There is the initial difficulty of ascribing to the state, or to any other group person, love, hate, jealousy and other intimate emotions which play a large part in individual morality. It seems plainly incongruous to say, as an eighteenth-century writer said, that 'a nation must love other nations as itself'.[19] For this reason, it is sometimes argued that the morality of the state must be confined to that formal kind of morality which can be codified in a set of rules and approximates to law, and that it cannot include such essentially personal qualities as altruism, generosity and compassion, whose obligations can never be precisely and rigidly defined. The state, like a public corporation, can – it is commonly said – be just, but not generous. This does not seem to be entirely true. We have already noted that group persons are commonly assumed to have moral as well as legal rights and obligations. When a bank or a public company subscribes to a Lord Mayor's Fund for assistance to victims of some great disaster, the act of generosity must be attributed not to the directors, whose pockets are not affected, and not to the shareholders, who are neither consulted nor informed, but to the bank or company itself. When the Treasury makes a 'compassionate grant'

in some case of hardship, the act of compassion is performed not by the official who takes the decision, and not by the Chancellor of the Exchequer in his individual capacity, but by the state. Some people expected 'the United States' to remit the debts owing to them from European states after the first world war, and criticized their refusal to do so on moral grounds. In other words, paradoxical as it may appear, we do, in certain circumstances, expect states and other group persons, not merely to comply with their formal obligations, but to behave generously and compassionately. And it is precisely this expectation which produces moral behaviour on behalf of a fictitious entity like a bank or a state. Banks subscribe to charitable funds and states make compassionate grants because public opinion expects it of them. The moral impulse may be traced back to individuals. But the moral act is the act of the group person.

Nevertheless, while most people accept the hypothesis that group persons have in certain conditions a moral duty to act altruistically as well as justly, the duty of the group person appears by common consent to be more limited by self-interest than the duty of the individual. In theory, the individual who sacrifices his interests or even his life for the good of others is morally praiseworthy, though this duty might be limited by duty to family or dependents. The group person is not commonly expected to indulge in altruism at the cost of any serious sacrifice of its interests. A bank or public company which failed to pay dividends owing to generous contributions to charities would probably be thought worthy of censure rather than praise. In his presidential campaign of 1932, Franklin Roosevelt referred tauntingly to Mr Hoover's reputation for humanitarian activities in Europe, and invited him to 'turn his eyes from his so-called "backward and crippled countries" to the great and stricken markets of Kansas, Nebraska, Iowa, Wisconsin and other agricultural states'.[20] It is not the ordinarily accepted moral duty of a state to lower the standard of living of its citizens by throwing open its frontiers to an unlimited number of foreign refugees, though it may be its duty to admit as large a number as is compatible with the interests of its own people. British supporters of the League of Nations who urged Great Britain to render assistance to victims of 'aggression' did not maintain that she should do this even to the detriment of her vital interests; they argued that she should render the assistance which she could reasonably afford[21] (just as a bank can reasonably afford to give 500 guineas to the victims of an earthquake). The accepted standard of international morality in regard to the altruistic virtues appears to be that a state should indulge in them in so far as this is not seriously incompatible with its more important interests. The result is that secure and wealthy groups can better afford to behave altruistically than groups which are continually preoccupied with the problem of their own security and solvency; and this circumstance provides such basis as there is for the assumption commonly made by Englishmen and Americans that the policies of their countries are morally more enlightened than those of other countries.

(2) It is, however, not merely true that the ordinary man does not demand from the group person certain kinds of moral behaviour which are demanded from the individual; he expects from the group person certain kinds of behaviour which he would definitely regard as immoral in the individual. The group is not only exempt from some of the moral obligations of the individual, but is definitely associated with pugnacity and self-assertion, which become positive virtues of the group person. The individual seeks strength through combination with others in the group; and his 'devotion to his community always means the expression of a transferred egoism as well as of altruism'.[22] If he is strong, he converts the group to the pursuit of his own ends. If he is weak, he finds compensation for his own lack of power to assert himself in the vicarious self-assertion of the group. If we cannot win ourselves, we want our side to win. Loyalty to the group comes to be regarded as a cardinal virtue of the individual, and may require him to condone behaviour by the group person which he would condemn in himself. It becomes a moral duty to promote the welfare, and further the interests, of the group as a whole; and this duty tends to eclipse duty to a wider community. Acts which would be immoral in the individual may become virtue when performed on behalf of the group person. 'If we were to do for ourselves what we are doing for Italy,' said Cavour to D'Azeglio, 'we should be great rogues.'[23] The same could truthfully have been said by many directors of public companies and promoters of good causes. 'There is an increasing tendency among modern men', writes Dr Niebuhr, 'to imagine themselves ethical because they have delegated their vices to larger and larger groups.'[24] In the same way we delegate our animosities. It is easier for 'England' to hate 'Germany' than for individual Englishmen to hate individual Germans. It is easier to be anti-Semitic than to hate individual Jews. We condemn such emotions in ourselves as individuals, but indulge them without scruple in our capacity as members of a group.

(3) These considerations apply in some measure to all group persons, though they apply with particular force to the state. There are, however, other respects in which we do not ordinarily demand from the state even the same standard of moral behaviour which we demand from other group persons. The state makes an altogether different kind of emotional appeal to its members from that of any other group person. It covers a far larger field of human activities, and demands from the individual a far more intensive loyalty and far graver sacrifices. The good of the state

comes more easily to be regarded as a moral end in itself. If we are asked to die for our country, we must at least be allowed to believe that our country's good is the most important thing in the world. The state thus comes to be regarded as having a right of self-preservation which overrides moral obligation. In the *Cambridge History of British Foreign Policy* published after the war, Professor Holland Rose condones the 'discreditable episode' of the seizure of the Danish fleet at Copenhagen in 1807 on the ground of Canning's belief that 'the very existence of Great Britain was at stake'.[25] Those who take a different view commonly argue that Canning was mistaken, not that he should have acted otherwise if his belief had been correct.

Other differences between the standards of morality commonly expected of the state and of other group persons arise from the fact that the state is the repository of political power and that there is no authority above the state capable of imposing moral behaviour on it, as a certain minimum of moral behaviour is imposed on other group persons by the state. One corollary of this is that we are bound to concede to the state a right of self-help in remedying its just grievances. Another corollary is the difficulty of securing the observance by all of a common standard; for while some moral obligations are always thought of as absolute, there is a strong tendency to make the imperativeness of moral obligations dependent on a reasonable expectation of the performance of the same duty by others. Conventions play an important part in all morality; and the essence of a convention is that it is binding so long as other people in fact abide by it. Barclays Bank or Imperial Chemicals Limited would incur moral censure if they employed secret agents to steal confidential documents from the safes of rival institutions, since such methods are not habitually employed by public companies against one another. But no stigma attaches to 'Great Britain' or 'Germany' for acting in this manner; for such practices are believed to be common to all the Great Powers, and a state which did not resort to them might find itself at a disadvantage. Spinoza argued that states would do likewise if it suited their interest.[26] One reason why a higher standard of morality is not expected of states is because states in fact frequently fail to behave morally and because there are no means of compelling them to do so.

(4) This brings us to the most fundamental difficulty which confronts us in our analysis of the moral obligations currently attributed to the state. It is commonly accepted that the morality of group persons can only be social morality (a state or a limited liability company cannot be a saint or a mystic); and social morality implies duty to fellow members of a community, whether that community be a family, a church, a club, a nation or humanity itself. 'No individual can make a conscience for

himself,' writes T. H. Green; 'he always needs a society to make it for him.'[27] In what sense can we find a basis for international morality by positing a society of states?

Is there an international community?

Those who deny the possibility of an international morality naturally contest the existence of an international community. The English Hegelian Bosanquet, who may be taken as a typical representative of this view, argues that 'the nation state is the widest organization which has the common experience necessary to found a common life',[28] and rejects with emphasis 'the assumption that humanity is a real corporate being, an object of devotion and a guide to moral duty'.[29] The reply to this would appear to be that a corporate being is never 'real' except as a working hypothesis, and that whether a given corporate being is an object of devotion and a guide to moral duty is a question of fact which must be settled by observation and not by theory, and which may be answered differently at different times and places. It has already been shown that there is in fact a widespread assumption of the existence of a worldwide community of which states are the units and that the conception of the moral obligations of states is closely bound up with this assumption. There is a world community for the reason (and for no other) that people talk, and within certain limits behave, as if there were a world community. There is a world community because, as Señor de Madariaga put it, 'we have smuggled that truth into our store of spiritual thinking without preliminary discussion'.[30]

On the other hand, it would be a dangerous illusion to suppose that this hypothetical world community possesses the unity and coherence of communities of more limited size up to and including the state. If we examine the ways in which the world community falls short of this standard of coherence, we shall have a clue to the underlying reasons for the shortcomings of international morality. It falls short mainly in two ways: (i) the principle of equality between members of the community is not applied, and is indeed not easily applicable, in the world community, and (ii) the principle that the good of the whole takes precedence over the good of the part, which is a postulate of any fully integrated community, is not generally accepted.

The principle of equality

(i) The principle of equality within a community is difficult to define. Equality is never absolute, and may perhaps be defined as an absence of discrimination for reasons which are felt to be irrelevant. In Great Britain, the reasons for which some receive higher incomes or pay more taxes than others are (rightly or wrongly) felt to be relevant even by most of those in the less-favoured categories, and the principle of equality is not therefore infringed. But the principle would be infringed, and the community broken, if people with blue eyes were less favourably treated than people with brown, or people from Surrey than people from Hampshire. In many countries, minorities *are* discriminated against on grounds which they feel to be irrelevant, and these minorities cease to feel, and to be regarded, as members of the community.[31]

In the international community such discrimination is endemic. It arises in the first place from the attitude of individuals. Gladstone is said on one occasion to have exhorted an audience of his fellow-countrymen to 'remember that the sanctity of life in the villages of the Afghan mountains among the winter snows is no less inviolable in the eyes of the Almighty than your own'.[32] It may safely be said that the eyes of the Almighty are not in this respect those of the great majority of Englishmen. Most men's sense of common interest and obligation is keener in respect of family and friends than in respect of others of their fellow-countrymen, and keener in respect of their fellow-countrymen than of other people. Family and friends form a 'face-to-face' group, between whom the sense of moral obligation is most likely to be strong. The members of a modern nation are enabled, through a more or less uniform education, a popular national press, broadcasting and travel facilities, and a skilful use of symbols,[33] to acquire something of the character of a 'face-to-face' group. The ordinary Englishman carries in his mind a generalized picture of the behaviour, daily life, thoughts and interests of other Englishmen, whereas he has no such picture at all of the Greek or the Lithuanian. Moreover, the vividness of his picture of 'foreigners' will commonly vary in relation to geographical, racial and linguistic proximity, so that the ordinary Englishman will be likely to feel that he has something, however slight, in common with the German or the Australian and nothing at all in common with the Chinese or the Turk.[34] An American newspaper correspondent in Europe is said to have laid down the rule that an accident was worth reporting if it involved the death of one American, five Englishmen, or ten Europeans. We all apply, consciously or unconsciously, some such standard of relative values. 'If it was not that China was so far away,' said Neville Chamberlain in the House of Commons on the occasion of Japanese bombing of Chinese cities, 'and that the scenes which were taking place there were so remote from our everyday consciousness, the sentiments of pity, horror and indignation which would be aroused by a full observation of those events might drive this people to courses which perhaps they had never yet contemplated.'[35] The same *motif* recurred in his national broadcast during the Czecho-Slovak crisis on 27 September 1938; 'How horrible, fantastic, incredible it is that we should be digging trenches and trying on gas-masks here because of a quarrel in a far-away country between people of whom we know nothing.'[36] These words were criticized in many quarters. But there is little doubt that they represented the initial reaction of the ordinary Englishman. Our normal attitude to foreigners is a complete negation of that absence of discrimination on irrelevant grounds which we have recognized as the principle of equality.

This attitude of the individual is reflected in the attitude of states to one another; and the difficulty is intensified by the structure of the international community. Even if equality between individuals of different countries were recognized, the inequalities between states would be none the less flagrant. The existing inequalities among a handful of known states subject to no external control are infinitely more glaring, more permanent and more difficult to forget than inequalities between the anonymous mass of citizens subject, at any rate in name, to the same law. The importance attached to the idea of equality in international politics is shown by the number and insistence of the demands based on it. 'Most-favoured-nation treatment', the 'Open Door', 'freedom of the seas', the Japanese claim for the recognition of racial equality in the Covenant of the League of Nations, the old German claim to 'a place in the sun', the more recent German claim to *Gleichberechtigung* or 'equality of status', have all been demands for the application of the principle of equality. The praises of equality were repeatedly sung in the Assemblies and Committees of the League of Nations – mainly, if not exclusively, by delegates of minor Powers.[37] Yet there is little attempt at consistency in the use of the term. Sometimes it merely means formal equality of states before the law. In other contexts, it may mean equality of rights, or equality of opportunity or equality of possessions. Sometimes it seems to mean equality between Great Powers. When Hitler argued that 'according to all common sense, logic and the general principles of high human justice ... all peoples ought to have an equal share of the goods of the world',[38] he hardly intended to convey that Lithuania ought to enjoy as much of 'the goods of the world' as Germany. Yet if we assume that equality of rights or privileges means proportionate, not absolute, equality, we are little advanced so long as there is no recognized criterion for determining the proportion. Nor would even this help us much. The trouble is not that Guatemala's rights and privileges are only proportionately, not absolutely, equal to those of the United States, but that such rights and privileges as Guatemala has are enjoyed only by the good will of the United States. The constant intrusion, or potential intrusion, of power renders almost meaningless any conception of equality between members of the international community.

The good of the whole and the good of the part

(ii) The other capital shortcoming of the international community is failure to secure general acceptance of the postulate that the good of the whole takes precedence over the good of the part. Great Britain possesses a common national consciousness because the man from Surrey will normally act on the assumption that the good of Great Britain is more important than the good of Surrey. One of the chief obstacles to the growth of a common German national consciousness was the difficulty in persuading Prussians, Saxons and Bavarians to treat the good of Germany as more important than the good of Prussia, Saxony and Bavaria. Now it is clear that, despite pious aspirations, people still hesitate to act on the belief that the good of the world at large is greater than the good of their own country. Loyalty to a world community is not yet powerful enough to create an international morality which will override vital national interests. Yet the conception of a community implies recognition of its good as something which its members are under an obligation to promote, and the conception of morality implies the recognition of principles of a universally binding character. If we refuse altogether to recognize the overriding claims of the whole, can any world community or any kind of international morality be said to exist at all?

This is the fundamental dilemma of international morality. On the one hand, we find the almost universal recognition of an international morality involving a sense of obligation to an international community or to humanity as a whole. On the other hand, we find an almost equally universal reluctance to admit that, in this international community, the good of the part (i.e. our own country) can be less important than the good of the whole. This dilemma is, in practice, resolved in two different ways. The first is the method, which Hitler borrowed from the Darwinian school, of identifying the good of the whole with the good of the fittest. The fittest are by assumption 'the bearers of a higher ethic';[39] and it is only necessary to prove in action that one's country is the fittest in order to establish the identity of its good with the good of the whole. The other method is that of the neo-liberal doctrine of the harmony of interests, of which Woodrow Wilson, Lord Cecil and Professor Toynbee have been quoted as representatives. This doctrine, like every doctrine of a natural harmony of interests, identifies the good of the whole with the security of those in possession. When Woodrow Wilson declared that American principles were the principles of mankind, or Professor Toynbee that the security of the British Empire was 'the supreme interest of the whole world',* they were in effect making the same claim made by Hitler that their countrymen are 'the

*See pp. 72, 74.

bearers of a higher ethic'; and the same result is produced of identifying the good of the whole international community with the good of that part of it in which we are particularly interested. Both these methods are equally fatal to any effective conception of international morality.

There is no escape from the fundamental dilemma that every community, and every code of morality, postulates some recognition that the good of the part may have to be sacrificed to the good of the whole. The more explicitly we face this issue in the international community, the nearer we shall be to a solution of our problem. The analogy of the national community, though imperfect, is once more helpful. Modern liberalism, wrote Hobhouse shortly before 1914, 'postulates, not that there is an actually existing harmony requiring nothing but prudence and judgement for its effective operation, but only that there is a possible ethical harmony to which ... men might attain, and that in such attainment lies the social ideal'.[40] The word 'ethical' betrays the break in the argument. The nineteenth-century 'harmony requiring nothing but prudence and judgement for its effective operation' was a harmony of interests. The 'ethical harmony' is one achieved by the sacrifice of interests, which is necessary precisely because no natural harmony of interests exists. In the national community, appeals to self-sacrifice are constantly and successfully made, even when the sacrifice asked for is the sacrifice of life. But even in the national community, it would be erroneous to suppose that the so-called 'harmony' is established solely through voluntary self-sacrifice. The sacrifice required is frequently a forced one, and the 'harmony' is based on the realistic consideration that it is in the 'interest' of the individual to sacrifice voluntarily what would otherwise be taken from him by force. Harmony in the national order is achieved by this blend of morality and power.

In the international order, the role of power is greater and that of morality less. When self-sacrifice is attributed to an individual, the sacrifice may or may not be purely voluntary. When self-sacrifice is attributed to a state, the chances are greater that this alleged self-sacrifice will turn out on inspection to be a forced submission to a stronger power. Yet even in international relations, self-sacrifice is not altogether unknown. Many concessions made by Great Britain to the Dominions cannot be explained in terms either of British interests or of submission to the stronger. Concessions made by Great Britain to Germany in the nineteen-twenties, ineffective as they were, were dictated, not wholly by British interests or by fear of Germany's strength, but by a belief in some conception of international morality which was independent of British interests. Any international moral order must rest on some hegemony of power. But this hegemony, like the supremacy of a ruling class within the state, is in itself a challenge to those who do not share it; and it must, if it is to survive, contain an element of give and take, of self-sacrifice on the part of those who have, which will render it tolerable to the other members of the world community. It is through this process of give and take, of willingness not to insist on all the prerogatives of power, that morality finds its surest foothold in international – and perhaps also in national – politics. It is, no doubt, useless to begin by expecting far-reaching sacrifices. The standard of what we can reasonably afford must not be pitched too high. But the course most detrimental to international morality is surely to pretend that the German people are the bearers of a higher ethic, or that American principles are the principles of humanity, or that the security of Great Britain is the supreme good of the world, so that no sacrifices at all by one's own nation are in fact necessary. When Professor Zimmern urges 'the ordinary man' to '*enlarge* his vision so as to bear in mind that the *public affairs* of the twentieth century are *world affairs*',[41] the most concrete meaning which can be given to this injunction is that the recognition of the principle of self-sacrifice, which is commonly supposed to stop short at the national frontier, should be extended beyond it. It is not certain that ordinary man will remain deaf to such an appeal. If the Chancellor of the Exchequer were to attempt to justify an increase in the income tax on the ground that it would make us better off, we should dismiss him as a humbug; and this is the kind of argument which is almost invariably used to justify any international policy involving apparent sacrifice of interests. A direct appeal to the need of self-sacrifice for a common good might sometimes prove more effective.

But it is necessary to clear up a further point on which many illusions are current. In the national community, we assume that in this process of self-sacrifice and give and take the giving must come principally from those who profit most by the existing order. In the international community, the assumption is commonly made by statesmen and writers of the satisfied Powers that the process of give and take operates only within the existing order and that sacrifices should be made by all to maintain that order. International peace, said Mr Eden once, must be 'based on an international order with the nations leagued together to preserve it'; and to this international peace 'each nation makes its own contribution because it recognizes that therein lies its own enduring interest'.[42] The fallacy latent in this and many similar pronouncements is fatal to any workable conception of international morality. The process of give and take must apply to challenges to the existing order. Those who profit most by that order can in the long run only hope to maintain it by making sufficient concessions to make it tolerable to those who profit by it least; and the responsibility for seeing that these changes take place as far as possible in an orderly way rests as much on the defenders as on the challengers. This leads us to an examination of the problems of law and change in international politics.

Notes

1. The Allied Governments in the Versailles Treaty attempted to revive this historic assumption by holding the ex-Kaiser personally responsible for acts of state; but the attempt was almost universally condemned as soon as passions began to cool. Modern dictatorships, however, helped to bring this conception back to fashion. Thus Professor Toynbee called the invasion of Abyssinia 'Signor Mussolini's deliberate personal sin' (*Survey of International Affairs, 1935*, ii. p. 3), though he would probably have felt it incongruous to describe the Hoare–Laval Plan as the 'personal sin' of Sir S. Hoare or Laval.
2. Duguit, for example, calls it 'valueless and meaningless anthropomorphism' (*Traité de droit constitutionnel*, i. ch. v).
3. Hall, *International Law* (8th ed.), p. 50; Pearce Higgins, *International Law and Relations*, p. 38.
4. This does not, of course, mean that the state is a necessary form of political organization, but only that, so long as the state *is* the accepted form, its personification is a necessary fiction. The same would apply to any other form (e.g. the class). The personification of the proletariat has gone far in Soviet Russia (e.g. the fiction that it 'owns' the means of production).
5. A striking example of confused thinking on this subject occurred in a recent letter to *The Times*. Commenting on the alleged British obligation to France in 1914, a distinguished professor of history wrote that 'Grey may have regarded his personal honour as involved in support of France, but he certainly did not think that of the Cabinet was' (*The Times*, 28 February 1939). The promise, if any, to support France must have been given by Grey not on his own behalf, but on behalf of Great Britain. Unless he believed that the whole Cabinet was under the same obligation as himself to see that Great Britain's promise was honoured, he could not properly have given it at all.
6. Quoted in *Democracy and War*, ed. G. E. C. Catlin, p. 128.
7. John Bright, *Speeches on Questions of Public Policy*, p. 479.
8. *Public Papers of Woodrow Wilson: War and Peace*, i. p. II.
9. *Intimate Papers of Colonel House*, ed. C. Seymour, iv. p. 28.
10. *League of Nations: Fourth Assembly*, i. p. 144.
11. *International Conciliation*, No. 334, p. 713.
12. Speech in the Reichstag, 21 May 1935.
13. The rules of war have since 1914 been exposed to an exacting test. The distinction between combatant and non-combatant grows less and less. A deliberate attack on so-called non-combatants may in fact promote important military objectives; and the conception of *unnecessary* suffering, which the belligerent is not entitled to inflict because it is not essential to his military

purpose, becomes more and more restricted and difficult to sustain. In short, modern conditions of warfare are doing much to break down, in one important point, a previously existing and effective sense of universal obligation.

14. Half a dozen synonyms, used quite indiscriminately, are quoted from recent documents by Dr G. Schwarzenberger (*American Journal of International Law*, xxxiii. p. 59). There is no reason to suspect sarcasm in the reference, in a Japanese Imperial Rescript of 1933, to 'the fraternity of nations'.

15. *Völkischer Beobachter*, 1 April 1939.

16. *The Times*, 9 May 1938.

17. *Foreign Affairs*, 15 March 1923, p. 95.

18. Zimmern, *Towards a National Policy*, p. 137.

19. Christian Wolff, quoted in H. Kraus, *Staatsethos*, p. 187.

20. Speech at the Metropolitan Opera House, New York, reported in the *New York Times*, 4 November 1932.

21. The League of Nations Union 'advocates sanctions only in cases where the number and resources of the governments co-operating on the League's behalf make it reasonably certain that the would-be aggressor will abandon his intention, so that war will not break out at all' (*Headway*, December 1937, p. 232).

22. R. Niebuhr, *Moral Man and Immoral Society*, p. 40.

23. Quoted in E. L. Woodward, *Three Studies in European Conservatism*, p. 297.

24. R. Niebuhr, *Atlantic Monthly*, 1927, p. 639.

25. *Cambridge History of British Foreign Policy*, i. pp. 363–4.

26. Spinoza, *Tractatus Politicus*, iii. § 14.

27. T. H. Green, *Prolegomena to Ethics*, p. 351.

28. B. Bosanquet, *The Philosophical Theory of the State*, p. 320.

29. B. Bosanquet, *Social and International Ideals*, p. 292.

30. S. de Madariaga, *The World's Design*, p. 3.

31. It is only in recent times that there has begun to be even a presumption that all inhabitants of a territory are members of the community. Like Jews in Nazi Germany, the coloured inhabitants of the Union of South Africa are today not regarded as members of the community. In the United States, most white Southerners would hesitate to admit that the negroes are members of the community in the same sense as they are themselves.

32. Quoted by the Delegate of Haiti in *League of Nations: Fifteenth Assembly*, 6th Committee, p. 43.

33. 'Moral attitudes always develop most sensitively in person-to-person relationships. That is one reason why more inclusive loyalties, naturally more abstract than immediate ones, lose some of their power over the human heart; and why a shrewd society attempts to restore that power by making a person the symbol of the community' (R. Niebuhr, *Moral Man and Immoral Society*, pp. 52–3).

34. The variations of feeling are naturally also influenced by current political prejudices.

35. House of Commons, 21 June 1938: *Official Report*, col. 936. A correspondent in *The Times*, commenting on 'the inconsistencies of compassion' in the international sphere, enquires whether 'the world's conscience' regards '100 dead or destitute Chinese as equivalent to one persecuted Jew', or whether it is 'simply that the Jews are near at hand, while the Chinese are a very long way away, and yellow at that' (*The Times*, 25 November 1938).

36. N. Chamberlain, *The Struggle for Peace*, p. 275.

37. Of the Great Powers only France, largely dependent for her position on the support of minor Powers, consistently advocated the principle of equality. 'There is not, and we trust there never will be,' said M. Blum on one occasion (*League of Nations: Sixteenth Assembly*, Part II, p. 28), 'an order of precedence among the Powers forming the international community. Were a hierarchy of States to be established within the League of Nations ... then the League would be ruined, both morally and materially' – a remarkable statement in view of the hierarchical constitution of the Council.

38. Speech in the Reichstag of 28 April 1939.

39. Hitler, *Mein Kampf*, p. 421.

40. L. T. Hobhouse, *Liberalism*, p. 129.

41. Zimmern, *The Prospects of Civilisation*, p. 26.

42. Anthony Eden, *Foreign Affairs*, p. 197.

Part Four

Law and Change

CHAPTER TEN

The Foundations of Law

No topic has been the subject of more confusion in contemporary thought about international problems than the relationship between politics and law. There is, among many people interested in international affairs, a strong inclination to treat law as something independent of, and ethically superior to, politics. 'The moral force of law' is contrasted with the implicitly immoral methods of politics. We are exhorted to establish 'the rule of law', to maintain 'international law and order' or to 'defend international law'; and the assumption is made that, by so doing, we shall transfer our differences from the turbulent political atmosphere of self-interest to the purer, serener air of impartial justice. Before adhering to these popular conceptions, we must examine rather carefully the nature and function of law in the international community and its relation to international politics.

The nature of international law

International law differs from the municipal law of modern states in being the law of an undeveloped and not fully integrated community. It lacks three institutions which are essential parts of any developed system of municipal law: a judicature, an executive and a legislature.

(1) International law recognizes no court competent to give on any issue of law or fact decisions recognized as binding by the community as a whole. It has long been the habit of some states to make special agreements to submit particular disputes to an international court for judicial settlement. The Permanent Court of International Justice, set up under the Covenant of the League, represents an attempt to extend and generalize this habit. But the institution of the Court has not changed international law: it has merely created certain special obligations for states willing to accept them.

(2) International law has no agents competent to enforce observance of the law. In certain cases, it does indeed recognize the right of an aggrieved party, where a breach of the law has occurred, to take reprisals against the offender. But this is the recognition of a right of self-help, not the enforce-

ment of a penalty by an agent of the law. The measures contemplated in Article 16 of the Covenant of the League, in so far as they can be regarded as punitive and not merely preventive, fall within this category.

(3) Of the two main sources of law – custom and legislation – international law knows only the former, resembling in this respect the law of all primitive communities. To trace the stages by which a certain kind of action or behaviour, from being customary, comes to be recognized as obligatory on all members of the community is the task of the social psychologist rather than of the jurist. But it is by some such process that international law has come into being. In advanced communities, the other source of law – direct legislation – is more prolific, and could not possibly be dispensed with in any modern state. So serious does this lack of international legislation appear that, in the view of some authorities, states do on certain occasions constitute themselves a legislative body, and many multilateral agreements between states are in fact 'lawmaking treaties' (*traités-lois*).[1] This view is open to grave objections. A treaty, whatever its scope and content, lacks the essential quality of law: it is not automatically and unconditionally applicable to all members of the community whether they assent to it or not. Attempts have been made from time to time to embody customary international law in multilateral treaties between states. But the value of such attempts has been largely nullified by the fact that no treaty can bind a state which has not accepted it. The Hague Conventions of 1907 on the rules of war are sometimes treated as an example of international legislation. But these conventions were not only not binding on states which were not parties to them, but were not binding on the parties *vis-à-vis* states which were not parties. The Kellogg–Briand Pact is not, as is sometimes loosely said, a legislative act prohibiting war. It is an agreement between a large number of states 'to renounce war as an instrument of national policy in their relations with one another'. International agreements are contracts concluded by states with one another in their capacity as subjects of international law, and not laws created by states in the capacity of international legislators. International legislation does not yet exist.

These shortcomings of international law, serious as they are, do not however deprive it of the title to be considered as law, of which it has all the essential characteristics. In particular, the relation of law to politics will be found to be the same in the international as in the national sphere.

It has been observed that the fundamental question of political philosophy is why men allow themselves to be ruled. The corresponding question which lies at the root of jurisprudence is why men obey the law. Why is law regarded as binding? The answer cannot be obtained from the law itself any more than a proof of Euclid's postulates can be obtained from Euclid. Law proceeds on the assumption that the question has been satisfactorily disposed of. But it is a question which cannot be burked by those who seek to justify the 'rule of law'. It applies to international as well as to municipal law. In international law, it sometimes takes the form of the question whether, and on what grounds, treaties are binding. The legal answer to this question is that treaties are binding in international law, which includes the rule (subject to some reservations which will be discussed presently) that treaties must be kept. But what the questioner probably means to ask is: Why is international law, and with it the rule that treaties must be kept, binding, and should they be regarded as binding at all? These are not questions which can be answered by international law. It is the purpose of this chapter to enquire in what domain the answer to them should be sought, and what that answer should be.

In approaching the problem of the ultimate authority of law, we shall find the same fundamental divergence which we have traced in the field of politics between utopians, who think in terms of ethics, and realists, who think in terms of power. Among students of law, the utopians are commonly known as 'naturalists', who find the authority of law in natural law, and the realists as 'positivists', who find the authority of law in the will of states. The terminology tends to become blurred and fluctuating. Some utopians purport to reject natural law, and adopt some other standard such as reason, utility, 'objective right',[2] 'ultimate sense of right',[3] or a 'fundamental norm'. Conversely, some positivists such as Spinoza purport to accept natural law, but empty it of its meaning by virtually identifying it with the right of the stronger. Other positivists fly the colours of 'the historical school of law' or of 'the economic interpretation of law'. But the fundamental divergence remains between those who regard law primarily as a branch of ethics, and those who regard it primarily as a vehicle of power.

The naturalist view of law

The naturalist view of law, like the utopian view of politics, has a longer history behind it than the positivist or realist view. In primitive communities, law is bound up with religion and, until a fairly late stage of human development, always appears to emanate from a god or a divinely appointed lawgiver. The secular civilization of the Greeks divorced law from religion, but not from morality. Greek thinkers found in the conception of natural law a higher unwritten law from which man-made law derived its validity and by which it could be tested. The acceptance of Christianity by the Roman Empire restored divine authority. Natural law was for a time identified with divine law, and it was only at the Renaissance that it resumed its independent role as a non-theological ethical standard. As we have seen, the seventeenth and eighteenth centuries revived in a new form the identification of natural law with reason. 'Law in general', says Montesquieu, 'is human reason, inasmuch as it governs all the peoples of the earth.'[4] It was under these auspices that modern international law was created by Grotius and his successors to meet the needs of the new nation-states which had arisen on the ruins of the mediaeval world. International law was therefore by origin strongly utopian. This was necessary and inevitable. The new conventions which came more or less effectively to govern relations between states grew no doubt out of practical needs. But they could never have secured as wide an acceptance as they did if they had not been treated as binding in virtue of natural law and universal reason. But here we shall note the recurrence of a paradox which is also apparent in the political field. Where practice is least ethical, theory becomes most utopian. Owing to the more primitive state of development of the international community, morality plays a smaller effective role in the practice of international law than of municipal law. In theories of international law, utopia tends to predominate over reality to an extent unparalleled in other branches of jurisprudence. Moreover, this tendency is greatest at periods when anarchy is most prevalent in the practice of nations. During the nineteenth century, a comparatively orderly period in international affairs, international jurisprudence took on a realist complexion. Since 1919, natural law has resumed its sway, and theories of international law have become more markedly utopian than at any previous time.

The modern view of natural law differs, however, in one important respect from the view which prevailed down to the end of the eighteenth century. Prior to that time, natural law had always been conceived as something essentially static, a fixed and eternal standard of right which must, in the nature of things, be the same yesterday, to-day and for ever. The historical tendency of nineteenth-century thought, which at first threatened to eclipse natural law altogether, gave it a new direction; and towards the end of the century there emerged the new conception of 'natural law with a variable content'.[5] Natural law, in this interpretation, connotes no longer something external, fixed and invariable, but men's innate feeling at any given time or place for what 'just law' ought to be. This revised definition of natural law helps us a little. It gets over the old crux that slavery was at one time thought to be sanctioned, and at another time to be prohibited, by natural law, or that private property is in some places regarded as a natural right and in other places as an infringement of natural right. We are now asked to treat law as binding because it is an emanation not of some eternal ethical principle, but of the ethical principles of a given time and community. This is, at any rate, a part of the truth. The ethical character of the impulse which lies behind many rules of law, municipal and international, including the rule of international law that treaties should be kept, will not be denied by any reasonable person. The prevalence in most European languages of words which bestride the frontier between law and ethics betrays a widespread conviction of the close relationship between them.

Nevertheless, this explanation why law is regarded as binding will turn out, on further examination, to be inadequate and in some degree misleading. The main crux about natural law is not that people differ from time to time and from place to place about what particular rules it prescribes (for this crux might be surmounted by the 'variable' theory), but that natural law (or reason or 'objective right' or any of its other substitutes) can be just as easily invoked to incite disobedience to the law as to justify obedience to it. Natural law has always had two aspects and two uses. It can be invoked by conservatives to justify the existing order, as when the rights of the rulers or the rights of property are alleged to rest on natural law. It can equally be invoked by revolutionaries to justify rebellion against the existing order. There is in natural law an anarchic element which is the direct antithesis of law. Theories of law which seek the ultimate authority of law in its ethical content can explain only why good laws (or laws regarded as good at a given time and place) are regarded as binding. Yet there is a fairly general consensus of opinion which regards as binding even laws recognized as bad; and it may be doubted whether any community could long survive in which such an opinion did not prevail. It is commonly admitted that there *may* be a right or duty to disobey a bad law. But in such cases, a conflict is recognized to exist between two duties; and it is generally felt that only the most exceptional circumstances justify a decision in favour of the duty to disobey. No theory of law seems adequate which explains that law is regarded as binding because it conforms to natural law or because it is good.

The realist view of law

The positivist or realist view of law was first clearly and explicitly stated by Hobbes, who defined law as a command: *Ius est quod iussum est*. Law is thus divorced altogether from ethics. It may be oppressive or otherwise immoral. It is regarded as binding because there is an authority which enforces obedience to it. It is an expression of the will of the state, and is used by those who control the state as an instrument of coercion against those who oppose their power. The law is therefore the weapon of the stronger. That contradictory thinker Rousseau, who elsewhere treats law as the antithesis of despotism, has recorded this view in emphatic terms: 'The spirit of the laws of all countries is always to favour the strong against the weak and him that has against him that has not. This drawback is inevitable, and there are no exceptions to it.'[6] According to Marx, all law is a 'law of inequality'.[7] The principal contribution of Marxism to the problem is its insistence on the relativity of law. Law reflects not any fixed ethical standard, but the policy and interests of the dominant group in a given state at a given period. Law, as Lenin puts it, is 'the formulation, the registration of power relations' and 'an expression of the will of the ruling class'.[8] The realist view of the ultimate basis of law is well summed up by Professor Laski: 'Legal rules are always seeking to accomplish an end deemed desirable by some group of men, and it is only by constant formulation of what that end is that we can obtain a realistic jurisprudence.'[9]

The realist answer to the question why law is regarded as binding contains, like the 'naturalist' answer, a part of the truth. Some people do in fact obey some laws because lawbreaking will bring them into unwelcome contact with the police and the courts. But no community could survive if most of its members were law-abiding only through an ever-present fear of punishment. As Laud says, 'No laws can be binding if there be no conscience to them';[10] and there is plenty of evidence of the difficulty of enforcing laws which seriously offend the conscience of the community or of any considerable part of it. Law is regarded as binding because it represents the sense of right of the community: it is an instrument of the common good. Law is regarded as binding because it is enforced by the strong arm of authority: it can be, and often is, oppressive. Both these answers are true; and both of them are only half-truths.

Law as a function of political society

If then we wish to reconcile these contradictory and inadequate half-truths, and to find a single answer to the question why law is regarded as binding, we must seek it in the relationship of law to politics. Law is regarded as binding because, if it were not, political society could not exist and there could be no law. Law is not an abstraction. It 'can only exist within a social framework. . . . Where there is law, there must be a society within which it is operative.'[11] We need not dwell on the old controversy whether, as the positivists held, the state creates law, or as the naturalists held, law creates the state. It is sufficient to say that no political society can exist without law, and that law cannot exist except in a political society.[12] The point has been clearly put by a contemporary German writer:

> All law is always the expression of a community. Every legal community (*Rechtsgemeinschaft*) has a common view of law (*Recht*)

determined by its content. It is an impossible undertaking to seek to construct a legal community without such a common view, or to establish a legal community before a minimum common view about the content of the community's law has been attained.[13]

Politics and law are indissolubly intertwined; for the relations of man to man in society which are the subject-matter of the one are the subject-matter of the other. Law, like politics, is a meeting place for ethics and power.

The same is true of international law, which can have no existence except in so far as there is an international community which, on the basis of a 'minimum common view', recognizes it as binding. International law is a function of the political community of nations. Its defects are due, not to any technical shortcomings, but to the embryonic character of the community in which it functions. Just as international morality is weaker than national morality, so international law is necessarily weaker and poorer in content than the municipal law of a highly organized modern state. The tiny number of states forming the international community creates the same special problem in law as in ethics. The evolution of general rules equally applicable to all, which is the basis of the ethical element in law, becomes extremely difficult. Rules, however general in form, will be constantly found to be aimed at a particular state or group of states; and for this reason, if for no other, the power element is more predominant and more obvious in international than in municipal law, whose subjects are a large body of anonymous individuals. The same consideration makes international law more frankly political than other branches of law.

Once therefore it is understood that law is a function of a given political order, whose existence alone can make it binding, we can see the fallacy of the personification of law implicit in such popular phrases as 'the rule of law' or 'the government of laws and not of men'. The man in the street tends to personify law as something which, whether he approves it or not, he recognizes as binding on him; and this personification is as natural for everyday purposes as the personification of the state. It is, nevertheless, dangerous to clear thinking. Law cannot be self-contained; for the obligation to obey it must always rest on something outside itself. It is neither self-creating nor self-applying. 'There are men who govern,' says a Chinese philosopher, 'but there are no laws that govern.'[14] When Hegel finds the embodiment of the highest moral good in the state, we are entitled to ask, What state? or, better, Whose state? When modern writers on international politics find the highest moral good in the rule of law, we are equally entitled to ask, What law? and Whose law? The law is not an abstraction. It cannot be understood independently of the political foundation on which it rests and of the political interests which it serves.

We shall also have no difficulty in detecting the fallacy in the common illusion that law is more moral than politics. A transaction, by becoming legal, does not become moral. To pay a workman less than a living wage is not any more moral because the wage is fixed in a contract signed by the workman and valid in law. The annexations of French territory by Germany in 1871 and of German territory by the Allies in 1919 may have been moral or immoral. But they are not made any more moral by the fact that they were registered in treaties signed by the defeated Powers and valid in international law. It is not in itself any more moral to deprive Jews of their property by a law to that effect than simply to send storm troopers to evict them. The laws of the Medes and Persians were probably not conspicuously moral. If the law is 'always seeking to accomplish an end deemed desirable by some group of men', the ethical character of the law is obviously conditioned by that end. Political action can be, and often is, invoked to remedy immoral or oppressive law. The peculiar quality of law which makes it a necessity in every political society resides not in its subject-matter, nor in its ethical content, but in its stability. Law gives to society that element of fixity and regularity and continuity without which no coherent life is possible. It is the fundamental basis of organized political society that the rights and duties of citizens in relation both to one another and to the state should be defined by law. Law which is uncertain in its interpretation or capricious in its application fails to fulfil its essential function.

Stability and continuity are, however, not the only requisites of political life. Society cannot live by law alone, and law cannot be the supreme authority. The political arena is the scene of a more or less constant struggle between conservatives, who in a general way desire to maintain the existing legal situation, and radicals, who desire to change it in important respects; and conservatives, national and international, have the habit of posing as defenders of law and of decrying their opponents as assailants of it. In democracies, this struggle between conservatives and radicals is carried on openly in accordance with legal rules. But these rules are themselves the product of a pre-legal political agreement. Every system of law presupposes an initial political decision, whether explicit or implied, whether achieved by voting or by bargaining or by force, as to the authority entitled to make and unmake law. Behind all law there is this necessary political background. The ultimate authority of law derives from politics.

Notes

1. The Carnegie Endowment has, for example, given the title *International Legislation* to a collection published under its auspices of 'multipartite instruments of general interest'.
2. Duguit, *Traité de droit constitutionnel*, i. p. 16.
3. Krabbe, *The Modern Idea of the State* (Engl. transl.), p. 110.
4. Montesquieu, *Esprit des Lois*, Book I. ch. iii.
5. The phrase comes from Stammler, whose *Lehre von dem richtigen Recbte* (1902–7) has been translated into English under the title *The Theory of Justice*.
6. Rousseau, *Émile*, Book IV.
7. Marx and Engels, *Works* (Russian ed.), xv. p. 272.
8. Lenin, *Works* (2nd Russian ed.), xv. p. 330; xii, p. 288.
9. *Representative Opinions of Mr Justice Holmes*, ed. Laski, Introduction.
10. Laud, Sermon IV, *Works*, i. p. 112.
11. Zimmern, *International Affairs*, xvii. (January–February 1938), p. 12.
12. 'We shall no longer ask whether the state is prior to law, or law is prior to the state. We shall regard them both as inherent functions of the common life which is inseparable from the idea of man. They will both be primordial facts: they will both have been given, as seeds or germs, coevally with man himself: they will both appear, as developed fruits, simultaneously with one another and in virtue of one another' (Gierke, *Natural Law and the Theory of Society*, Engl. transl., p. 224).
13. F. Berber, *Sicherheit und Gerechtigkeit*, p. 145.
14. Hsun-tze, quoted in Liang Chi-chao, *History of Chinese Political Thought*, p. 137. A perfect illustration of the confusion which results from treating law as something self-contained and self-applying may be found in a reported dictum of Mr Winston Churchill: 'There must be the assurance that some august international tribunal shall be established which will uphold, enforce and itself obey the law' (*Manchester Guardian*, 12 December 1938). If Mr Churchill had paused to ask *who* would establish the august tribunal, *who* would enforce its decisions, *who* would make the law and *who* would see that the tribunal obeyed it, the implications of this apparently simple proposition would have become apparent.

CHAPTER ELEVEN

The Sanctity of Treaties

One of the functions of law necessary to civilized life is to protect rights which have been created by private contracts concluded in a manner recognized by the law as valid. International law upholds, with some reservations, rights created by international treaties and agreements. This principle is essential to the existence of any kind of international community and is, as we have seen, recognized in theory by all states. The fact that the only written obligations of states are those contained in treaties, and that customary international law is limited in scope and sometimes uncertain in content, has given to treaties a more prominent place in international law than is occupied by contracts in municipal law. Indeed the contents of treaties are sometimes misleadingly spoken of as if they were a part of international law itself, though nobody would regard the provisions of a contract between Smith and Robinson as a part of municipal law. The principle of the sanctity of treaties has thus been thrown into undue relief, which was further intensified by the controversy over the peace treaties of 1919–20. Between the two wars writers, especially those from countries interested in the maintenance of the peace settlement, attempted to treat the rule *pacta sunt servanda* not merely as a fundamental rule of international law, but as the cornerstone of international society – an attitude mockingly described by a German writer as '*pacta-sunt-servanda*-ism'.[1] The issue has become one of the most contentious in the whole field of international politics; and confusion has often been caused by failure to distinguish between 'the sanctity of treaties' as a rule of international law and 'the sanctity of treaties' as a principle of international ethics.

The legal and moral validity of treaties

In spite of the universal recognition by all countries that treaties are in principle legally binding, international law before 1914 was reluctant to treat as absolute the binding character of treaty obligations. Account had to be taken of the fact that, while states interested in maintenance of the *status quo* vigorously asserted the unconditional validity of treaties in international law, a state whose interests were adversely affected by a treaty commonly repudiated it as soon as it could do so with impunity. France in 1848 announced that 'the treaties of 1815 are no longer valid in the eyes of the French Republic'.[2] Russia in 1871 repudiated the Straits Convention placing restrictions on the passage of her warships which had been imposed on her at the conclusion of the Crimean War. These were merely the most conspicuous of several similar nineteenth-century occurrences. To meet such conditions, international lawyers evolved the doctrine that a so-called *clausula rebus sic stantibus* was implicit in every treaty, i.e. that the obligations of a treaty were binding in international law so long as the conditions prevailing at the time of the conclusion of the treaty continued, and no longer. This doctrine, if carried to its logical conclusion, would appear to lead to the position that a treaty had no authority other than the power relationship of the parties to it, and that when this relationship alters the treaty lapses. This position was not infrequently adopted. 'Every treaty', wrote Bismarck in a famous phrase, 'has the significance only of a constatation of a definite position in European affairs. The reserve *rebus sic stantibus* is always silently understood.'[3] The same effect is produced by the doctrine occasionally propounded that a state enjoys the unconditional right to denounce any treaty at any time. This view was stated in its most uncompromising form by Theodore Roosevelt: 'The nation has as a matter of course a right to abrogate a treaty in a solemn and official manner for what she regards as a sufficient cause, just exactly as she has a right to declare war or exercise another power for a sufficient cause.'[4] Woodrow Wilson observed in private conversation during the Peace Conference that, when he was a teacher of international law, he had always supposed that a state had the power to denounce any treaty by which it was bound at any time.[5] In 1915, a distinguished neutral international lawyer of the 'naturalist' school wrote of the rule *pacta sunt servanda* that 'nobody regards it as a rule of law which is valid without exception either within or without the state'.[6]

Even Great Britain which, as the strongest Power in the world, had most interest in upholding the validity of treaties, was manifestly disinclined to accept the view that treaty obligations were unconditionally binding. The most famous example is that of the Belgian Guarantee Treaty of 1839, under which the principal European Powers, including Great Britain, bound themselves jointly and severally to resist any violation of the neutrality of Belgium by one of their number. In 1870 Gladstone told the House of Commons, in a passage which was cited with approval by Grey in his speech of 3 August 1914, that he was 'not able to subscribe to the doctrine of those who have held in this House what plainly amounts to an assertion that the simple fact of the existence of the guarantee is binding on every party of it, irrespective altogether of the particular position in which it may find itself at the time that the occasion for acting on the guarantee arises'. Such an interpretation Gladstone thought 'rigid' and 'impracticable'.[7] A confidential minute written in 1908 by Lord Hardinge, then Permanent Under-Secretary of State for Foreign Affairs, was conceived in the same spirit:

> The liability undoubtedly exists . . . but whether we could be called on to carry out our obligation and to vindicate the neutrality of Belgium in opposing its violation must necessarily depend upon our policy at the time and the circumstances of the moment. Supposing that France violated the neutrality of Belgium in a war against Germany it is, under present circumstances, doubtful whether England or Russia would move a finger to maintain Belgian neutrality, while, if the neutrality of Belgium were violated by Germany, it is probable that the converse would be the case.

Grey, commenting in a further minute, merely observed that this reflexion was 'to the point'.[8]

Another principle not less elastic than the *clausula* has sometimes been invoked to justify non-fulfilment of international obligations – the principle of 'necessity' or 'vital interests'. It is a well-known legal maxim that nobody can be called on to perform the impossible; and the impossible is sometimes held in international law to include acts detrimental to the vital interests (meaning primarily the security) of the state. Some writers have specifically held that every state has a legal right of self-preservation which overrides any obligation to other states. This view is likely to carry particular weight in time of war. In its note of protest against British blockade measures in December 1914, the United States Government laid it down as the principle of international law that belligerents should not interfere with neutral commerce 'unless such interference is manifestly an imperative necessity to protect their national safety, and then only to the extent that it is a necessity'. The British Government gratefully accepted this interpretation, and was thenceforth able to justify its blockade activities on the uncontested ground of an 'imperative necessity' whose requirements nobody was as well qualified as itself to assess.[9] In such emergencies, the layman is apt to discard legal niceties and arrive at the same result by other methods. At the time of the Jameson Raid, *The Times* published a poem by the Poet Laureate which opened with these disarming lines:

> Let lawyers and statesmen addle
> Their pates over points of law:

If sound be our sword and saddle
And gun-gear, who cares one straw?[10]

'Damn the law, I want the Canal built,' was a saying popularly attributed to Theodore Roosevelt at the time of the Panama crisis. In 1939 a Japanese 'naval spokesman', commenting on the boarding of foreign vessels in Chinese waters by Japanese patrols, is reported to have said: 'It is not a question of having the right to do this. It is something which is necessary and we are doing it.'[11] 'Once it [i.e. the nation] is in danger of oppression or annihilation,' wrote Hitler, 'the question of legality plays a subordinate role.'[12]

Indeed, where justification is explicitly or implicitly offered for the nonfulfilment of treaty obligations, it is often difficult to discover from the words used whether the alleged justification is based on legal or on moral grounds. Is the view taken that, by the operation of *clausula rebus sic stantibus* or for some other reason, the obligation is no longer binding in law? Or is the legal obligation admitted, and is it argued that the state is entitled to disregard the law on the ground that it is immoral, unreasonable or impracticable, just as the citizen is sometimes morally entitled to disregard the national law? Broadly speaking, it may be said that prior to 1914 the rule *pacta sunt servanda* was elastically interpreted and the nonfulfilment of obligations was apt to be defended as legally admissible, whereas since 1919 the interpretation of the rule has tended to become more rigid, and nonfulfilment has been defended mainly on the ground that considerations of reason or morality entitled the state to disregard its strictly legal obligation. The dilemma of international law is that of ecclesiastical dogma. Elastic interpretation adapted to diverse needs increases the number of the faithful. Rigid interpretation, though theoretically desirable, provokes secessions from the church. It cannot be doubted that the more frequent and open repudiation of the rules of international law since 1919 has been due in part to the well-intentioned efforts of the victorious Powers to strengthen those rules and to interpret them with greater rigidity and precision.

An examination of the numerous breaches of treaty obligations during this period yields less definite results than might have been expected; for the state concerned in many cases defended itself either by denying that any breach of treaty obligations has occurred, or by alleging that the treaty had in the first instance been violated by the other party. In December 1932, the French Chamber of Deputies refused to carry out the French War debt agreement with the United States on the ground that 'the determining circumstances' had changed since the conclusion of the agreement six years earlier – the nearest approach since 1919 to an explicit invocation of the *clausula rebus sic stantibus*.[13] The British default on the Anglo-American War debt agreement was justified on the ground of 'economic necessity'. But the main ground of the argument was not legal, but moral: the burden imposed by the agreement was 'unreasonable' and 'inequitable'.[14] *The Times* took the view that the debt 'had not the same moral validity as an ordinary commercial transaction'.[15] At an earlier stage, Neville Chamberlain, then Chancellor of the Exchequer, had explicitly admitted that the obligation was legally binding, but had appealed to other obligations which might be rated higher than those of law:

> When we are told that contracts must be kept sacred, and that we must on no account depart from the obligations which we have undertaken, it must not be forgotten that we have other obligations and responsibilities, obligations not only to our countrymen, but to many millions of human beings throughout the world, whose happiness or misery may depend upon how far the fulfilment of these obligations is insisted upon on the one side and met on the other.[16]

In repudiating the military clauses of the Versailles Treaty in March 1935, Germany based her action on the alleged failure of the other parties to the treaty to implement their own obligations to disarm. A year later, the repudiation of the Locarno Treaty was justified on the ground that, through the action of France in concluding the Franco-Soviet Pact, the treaty had 'ceased in practice to exist'.[17] These were at any rate ostensibly legal arguments. But in a public speech shortly after the occupation of the Rhineland, Hitler rejected the legal in favour of the moral plea: 'If the rest of the world clings to the letter of treaties, I cling to an eternal morality.'[18]

On the whole, therefore, it may be said that breaches of treaties between the two wars were excused, not on the legal ground of derogations admitted by international law to the principle of the sanctity of treaties, but on the ethical ground that certain treaties, though legally binding, lack moral validity. It was not denied that breaches of such treaties are technical breaches of international law; but they were an offence against international morality. It is important for the student of international ethics and international law to study the qualities which were popularly supposed to make treaties morally disreputable and therefore morally invalid.

Treaties signed under duress

In the first place, it came to be felt that there was a moral taint about treaties signed under duress. This feeling attached itself mainly to the Versailles Treaty, signed by Germany under the duress of a five-day ultimatum. German propaganda worked hard to popularize the conception of the

Versailles Treaty as a *Diktat* which had no moral validity; and the idea enjoyed widespread currency after the conclusion of the Locarno Treaty, when British and French statesmen rashly vied with Stresemann in emphasizing the moral significance of the voluntary acceptance by Germany of some of the obligations accepted under duress at Versailles. The attitude adopted to treaties concluded under duress is dependent on the attitude adopted to war; for every treaty which brings a war to an end is almost inevitably accepted by the loser under duress. So long therefore as any kind of war whatever is recognized as moral, treaties concluded under duress cannot be unconditionally condemned as immoral. The moral objections most frequently expressed against the Versailles Treaty seem, in fact, to have been based not so much on its signature under duress as on the severity of its contents, and on the fact that the Allied Governments, reversing the procedure followed at all important peace conferences down to and including that of Brest-Litovsk, refused to engage in oral negotiations with the plenipotentiaries of the defeated Power. This act of unwisdom probably discredited the treaty more than the ultimatum which preceded its signature.

Inequitable treaties

Secondly, the view was commonly taken that treaties may be morally invalidated by the character of their contents. There cannot indeed be any rule of international *law* corresponding to the rule of municipal law voiding contracts which are 'immoral' or 'contrary to public policy'. The absence of an international political order makes impossible any *legal* definition of international public policy or of what is internationally immoral.[19] But those who regard the contents of a given international treaty as immoral will, generally speaking, concede to the injured state the moral right to repudiate it; for international law provides no other means of redress. It should, moreover, be observed that there is a tendency to concede the same moral right to repudiate a treaty which is not, properly speaking, immoral, but which is inequitable in the sense that it imposes conditions flagrantly incompatible with the existing relations of power between the contracting parties. The disarmament clauses of the Versailles Treaty were widely regarded as lacking in validity because it was unreasonable to impose a position of permanent inferiority on a Great Power. In general, the reproach was levelled against the Versailles Treaty that it sought to perpetuate the temporary weakness of Germany due to her collapse at the end of the War. This argument is not perhaps strictly ethical, since it is rooted in the power position and recognizes a moral right based simply on strength. But it is an illustration of the curious way in which power and ethics are intertwined in all political problems. A somewhat similar case arose in connexion with Article 16 of the Covenant of the League of Nations. When the United States failed to ratify the Covenant, it was widely felt that the obligations imposed by that Article were no longer morally binding, since members of the League could not reasonably be expected to take measures which might bring on them the enmity of so powerful a country. The test of what is commonly recognized as reasonable applies to the moral validity of treaties as to other problems of international morality.

Treaties as instruments of power

The third consideration which is sometimes invoked to deny the morally binding character of international treaties is of a more sweeping kind. It is designed to cast doubts on the moral credit not of particular treaties, but of all treaties as being by their nature instruments of power and therefore devoid of moral value. A Marxist writer has argued that, in capitalist society, the legal enforcement of contracts is merely a method of using the power of the state to protect and further the interests of the ruling class.[20] In the same way, it can be maintained with considerable show of reason that insistence on the legal validity of international treaties is a weapon used by the ruling nations to maintain their supremacy over weaker nations on whom the treaties have been imposed. Such an argument is implicit in the realist view of law as an oppressive instrument of power divorced from ethics.

The argument is assisted by the elastic and inconsistent manner in which the doctrine of the sanctity of treaties has been applied in the practice of states. In 1932–33, the French and British Governments were insisting with particular vehemence that the disarmament clauses of the Versailles Treaty were legally binding on Germany, and could be revised only with the consent of the interested Powers. In December 1932, the French Chamber of Deputies found reasons for refusing to carry out the French war debt agreement with the United States. In June 1933, the British Government ceased to pay the regular instalments due under its war debt agreement, substituting minor 'token payments'; and a year later these token payments came to an end. Yet in 1935 Great Britain and France once more joined in a solemn condemnation of Germany for unilaterally repudiating her obligations under the disarmament clauses of the Versailles Treaty. Such inconsistencies are so common that the realist finds little difficulty in reducing them to a simple rule. The element of power is inherent in every political treaty. The contents of such a treaty reflect in some degree the relative strength of the contracting parties. Stronger states will insist on the sanctity of the treaties concluded by them with weaker states. Weaker states will renounce treaties concluded by them with stronger states as soon as the power position alters and the weaker state feels itself strong enough to reject or modify the obligation. Since 1918, the United States have concluded no treaty with a stronger state, and have therefore unreservedly upheld the sanctity of treaties. Great Britain concluded the war debt agreement with a country financially stronger than herself, and defaulted. She concluded no other important treaty with a stronger Power and, with this single exception, upheld the sanctity of treaties. The countries which had concluded the largest number of treaties with states stronger than themselves, and subsequently strengthened their position, were Germany, Italy and Japan; and these are the countries which renounced or violated the largest number of treaties. But it would be rash to assume any *moral* distinction between these different attitudes. There is no reason to assume that these countries would insist any less strongly than Great Britain or the United States on the sanctity of treaties favourable to themselves concluded by them with weaker states.

The case is convincing as far as it goes. The rule *pacta sunt servanda* is not a moral principle, and its application cannot always be justified on ethical grounds. It is a rule of international law; and as such it not only is, but is universally recognized to be, necessary to the existence of an international society. But law does not purport to solve every political problem; and where it fails, the fault often lies with those who seek to put it to uses for which it was never intended. It is no reproach to law to describe it as a bulwark of the existing order. The essence of law is to promote stability and maintain the existing framework of society; and it is perfectly natural everywhere for conservatives to describe themselves as the party of law and order, and to denounce radicals as disturbers of the peace and enemies of the law. The history of every society reveals a strong tendency on the part of those who want important changes in the existing order to commit acts which are illegal or which can plausibly be denounced as such by conservatives. It is true that in highly organized societies, where legally constituted machinery exists for bringing about changes in the law, this tendency to illegal action is mitigated. But it is never removed altogether. Radicals are always more likely than conservatives to come into conflict with the law.

Before 1914, international law did not condemn as illegal resort to war for the purpose of changing the existing international order; and no legally constituted machinery existed for bringing about changes in any other way. After 1918 opinion condemning 'aggressive' war became almost universal, and nearly all the nations of the world signed a pact renouncing resort to war as an instrument of policy. While therefore resort to war for the purpose of altering the *status quo* now usually involves the breach of a treaty obligation and is accordingly illegal in international law, no effective international machinery has been constituted for bringing about changes by pacific means. The rude nineteenth-century system, or lack of system, was logical in recognizing as legal the one effective method of changing the *status quo*. The rejection of the traditional method as illegal and the failure to provide any effective alternative have made contemporary international law a bulwark of the existing order to an extent unknown in previous international law or in the municipal law of any civilized country. This is the most fundamental cause of the recent decline of respect for international law; and those who, in deploring the phenomenon, fail to recognize its origin, not unnaturally expose themselves to the charge of hypocrisy or of obtuseness.

Of all the considerations which render unlikely the general observance of the legal rule of the sanctity of treaties, and which provide a plausible moral justification for the repudiation of treaties, this last is by far the most important. Respect for international law and for the sanctity of treaties will not be increased by the sermons of those who, having most to gain from the maintenance of the existing order, insist most firmly on the morally binding character of the law. Respect for law and treaties will be maintained only in so far as the law recognizes effective political machinery through which it can itself be modified and superseded. There must be a clear recognition of that play of political forces which is antecedent to all law. Only when these forces are in stable equilibrium can the law perform its social function without becoming a tool in the hands of the defenders of the *status quo*. The achievement of this equilibrium is not a legal, but a political task.

Notes

1. Walz in *Deutsches Recht*, Jg. IV. (1934), p. 525. Professor Lauterpacht's remark that the rule *pacta sunt servanda* 'constitutes the highest, irreducible, final criterion' in international society (*The Function of Law in the International Community*, p. 418) is a good example of the attitude criticized.
2. Lamartine's Circular of 5 March 1848, published in the *Moniteur* of that date.
3. Bismarck, *Gedanken und Erinnerungen*, ii. p. 258.
4. Quoted in H. F. Pringle, *Theodore Roosevelt*, p. 309.
5. Miller, *The Drafting of the Covenant*, i. p. 293.
6. Krabbe, *The Modern Idea of the State* (Engl. transl.), p. 266.
7. Quoted in Grey, *Speeches on Foreign Affairs, 1904–1914*, p. 307.
8. *British Documents on the Origin of the War*, ed. Gooch and Temperley, viii. pp. 377–8.
9. The correspondence was published in Cmd 7816 of 1915.

10. *The Times*, 11 January 1896.
11. *The Times*, 26 May 1939.
12. Hitler, *Mein Kampf*, p. 104.
13. Resolution of 14 December 1932, in *Documents on International Affairs, 1932*, pp. 80–2.
14. The quotations are from the British note of 4 June 1934 (Cmd 4609).
15. *The Times*, 2 June 1934.
16. Speech in the House of Commons, 14 December 1932, in *Documents on International Affairs, 1932*, p. 128.
17. *Diplomatic Discussions Directed Towards a European Settlement*, Cmd 5143, p. 78.
18. Quoted in Toynbee, *Survey of International Affairs, 1936*, p. 319. Such pleas are not peculiarly modern and have often been regarded as legitimate. As recently as 1908, a distinguished English historian used of Pitt words which, with the bare change of proper names, are precisely apposite to Hitler's attitude: 'His support of the British claim as "from God and Nature" to override the artificial restrictions of unjust treaties, his denunciation of the Convention of the Pardo as "a stipulation for the national ignominy", voiced the inarticulate sentiment of the new England' (*Quarterly Review*, October 1908, p. 325). A later passage in the same article runs as follows: 'By the alchemy of his own intense vision and political ideals, he imposed on England a conception of national development and national ends based on an ideal of Imperialist expansion to realize which the nation must sacrifice everything or cease to believe in its own right and power to exist' (*ibid.*, pp. 334–5). It is interesting to observe that the writer clearly regarded these phrases as eulogistic.
19. Some German writers after 1919 tried to maintain that treaties are invalid in international law if they conflict with the 'natural law of nations'. The literature is reviewed by Verdross, *American Journal of International Law*, xxxi. (October 1937), pp. 571 *sqq*. But this view has found little support elsewhere. On the occasion of a judgement by the Permanent Court of International Justice in 1934, the German judge, in an individual opinion, expressed the view that the Court 'would never apply a convention whose contents were contrary to *bonnes mœurs*' (*Permanent Court of International Justice*, Series A/B No. 63, p. 150). But the Court as such never appears to have committed itself to this proposition.
20. Renner, *Die Rechtsinstitute des Privatrechts und ibre soziale Funktion*, p. 55.

CHAPTER TWELVE

The Judicial Settlement of International Disputes

Besides upholding legal rights, the law provides machinery for settling disputes about these rights. The jurisdiction of national courts is compulsory. Any person cited before a court must enter an appearance or lose his case by default; and the decision of the court is binding on all concerned.

International law, though it provides machinery for the settlement of disputes, recognizes no compulsory jurisdiction. Down to the end of the nineteenth century, the judicial process as applied to international disputes almost invariably took the form of an *ad hoc* agreement to submit a particular dispute to an arbitrator or arbitrators, whose method of appointment was fixed by the agreement and whose verdict was accepted in advance as binding. Under the Hague Convention of 1899, a Permanent Court of Arbitration was established at The Hague. This was, however, not a court, but a standing panel from which suitable arbitrators could be selected by states desiring to resort to arbitration. The Permanent Court of International Justice established under the Covenant of the League of Nations really was a court sitting as such. But it exercised jurisdiction only with the consent of the parties, whether that consent was expressed in an *ad hoc* agreement relating to the particular dispute or in a general agreement between the parties to submit to the Court all disputes falling within certain categories. 'It is well established in international law', declared the Court itself in one of its judgements, 'that no state can, without its consent, be compelled to submit its disputes with other states either to mediation or to arbitration or to any other kind of pacific settlement.'[1]

Justiciable and non-justiciable disputes

In municipal law, all disputes are theoretically justiciable; for if the point at issue is covered by no legal rule, the answer of the court will be that the complainant has no case. It is true that the complainant may not be satisfied with this answer, and may seek to obtain redress by political action. But this merely means that he does not want a legal answer, not that the law has no answer to give, or that the answer is not legally binding. In international law, all disputes are not justiciable; for no court is competent unless the parties to the dispute have agreed to confer jurisdiction on it and to recognize its decision as binding. Many treaties are in existence in which the parties define the kinds of disputes which they agree to recognize as justiciable as between themselves. In some treaties before 1914, disputes of certain limited and specific categories were recognized as justiciable. In others, the definition of justiciable disputes took a negative and somewhat elastic form: the parties to the treaty undertook to submit to arbitration any dispute between them which did not affect their 'vital interests', 'independence' or 'national honour'. The nearest approach to a definition of justiciable disputes was contained in Article 13 of the League Covenant, and repeated in Article 36 of the Permanent Court, which enumerated various kinds of dispute 'declared to be among those which are generally suitable for submission to arbitration or judicial settlement'. Finally several arbitration treaties concluded after 1919, notably those negotiated at Locarno, recognized as justiciable what were called disputes between the parties 'as to their respective rights'.

The formulae of the Covenant and the Statute and of the Locarno arbitration treaties have given a strong impetus to the idea that international disputes could be classified by an objective test as *ipso facto* justiciable and *ipso facto* non-justiciable. Any such classification rests on an illusion. The formulae in question provide no objective definition of a justiciable dispute. They merely indicate certain kinds of dispute which the parties to these instruments agree to recognize as justiciable between themselves. The formula of the Covenant and the Statue is not really a definition at all, but an enumeration of examples which does not purport to be either exhaustive or (as the qualification 'generally' shows) authoritative.[2] The Locarno formula is an attempt to give an objective character to the distinction between justiciable and non-justiciable disputes by identifying it with the distinction between conflicts of legal right and conflicts of interest. This formula has little practical value. It merely binds the parties to recognize as justiciable any dispute which they agree to regard as an issue of law. Either party can withdraw any dispute from arbitration by the simple process of placing itself on some other ground than that of legal right. Thus, the British Government, if it had been bound by such a treaty, would presumably have refused to submit to arbitration its default on the war debt agreement with the United States on the ground that the point at issue was not the legal right of the United States to demand payment, and that the dispute was not therefore one as to 'respective rights'. As Professor Lauterpacht has conclusively shown, there is no objective criterion of the 'suitability' of a dispute for judicial settlement. 'It is not the nature of an individual dispute which makes it unfit for judicial settlement but the unwillingness of a state to have it settled by the application of law.'[3] The question which confronts us is twofold: Why are states willing to submit only certain kinds of dispute to judicial settlement, and why do they find it so difficult to define in clear terms what kinds of dispute they are willing to submit?

The answer to this question must be sought in the necessary relation of law to politics. The judicial settlement of disputes presupposes the existence of law and the recognition that it is binding; and the agreement which makes the law and which treats it as binding is a political fact. The applicability of judicial procedure depends therefore on explicit or implicit political agreement. In international relations, political agreement tends to be restricted to those spheres which do not affect the security and existence of the state; and it is primarily in these spheres that the judicial settlement of disputes is effective. The majority of international disputes which have in the past been settled by arbitration or by some other legal procedure have been either pecuniary claims or disputes about national frontiers in remote and sparsely inhabited regions. The exclusion, in arbitration treaties concluded before 1914, of disputes affecting 'vital interests', 'independence' or 'national honour' meant the exclusion of precisely those matters on which political agreement could not be attained. When political disagreement threatened, arbitration was recognized as impracticable. We shall see presently that what is virtually the same reservation was maintained in subsequent agreements for arbitration or judicial settlement in the form of the exclusion from these agreements of disputes endangering the sanctity of existing treaties or existing legal rights.

The same consideration explains why no definition of disputes recognized as justiciable can be universally or permanently valid; for political agreement is a factor which varies from place to place and from time to time.[4] Prior to 1917 there was a general political understanding throughout the world that the property rights of individuals were valid, and that a foreigner whose property was for any reason confiscated by the government of the country in which it was situated had a claim in international law to compensation. So long as this understanding existed, claims based upon it could be settled by arbitration. With the establishment of the Soviet régime in Russia, this understanding ceased to apply to that country; and when the Soviet Government made its first important international appearance at the Genoa Conference in 1922, it was careful to scout in advance the idea that property claims against it should be submitted to arbitration. 'In the trial of disputes of this kind', ran the memorandum which it submitted to the Conference, 'the specific disagreements will inevitably end in opposing to one another two forms of property. ... In such circumstances there can be no question of an impartial super-arbiter.' And when, at the subsequent Hague Conference, the British delegate pathetically enquired 'whether it would be impossible to find a single impartial judge in the whole world', Mr Litvinov firmly replied that 'it was necessary to face the fact that there was not one world, but two, a Soviet world and a non-Soviet world'.[5] 'Impartiality' is a meaningless concept where there is no common ground at all between the two contending views. Judicial procedure cannot operate without accepted political postulates.

The assumption of the British delegate just quoted that the obstacle to international arbitration was the difficulty of finding impartial judges had been heard on previous occasions. 'The great obstacle to the extension of arbitration', declared the American delegate at the Hague Conference of 1907, 'is not the unwillingness of civilized nations to submit their disputes to the decision of an arbitral tribunal; it is rather an apprehension that the tribunal selected will not be impartial.' Lord Salisbury is quoted in a similar sense.[6] This opinion rests on a misapprehension. The potential personal bias of the international judge is not the real stumbling-block. The popular prejudice against submitting matters of national concern to the verdict of a 'foreigner' is based primarily, not on the belief that the foreign judge will be biased as between the parties, but on the fact that there are certain fundamentals of a political character which we are not prepared to have challenged by any foreign authority, whether judicial or political. The abolition of private ownership for Soviet Russia, the right of blockade for Great Britain, the Monroe Doctrine for the United States are familiar examples of such political fundamentals. Such fundamentals need not, however, be major issues at all. Palmerston treated the Don Pacifico episode in 1850, and Signor Mussolini the murder of an Italian general in Greece in 1923, as political issues which they were not prepared to submit to judicial settlement.[7]

But there is another and more general sense in which the absence of common political presupposition impedes the development of the judicial process in the international community. Municipal law, though far more fully and minutely developed than international law, is never wholly self-sufficing. The application of the law to the particular case is always liable to involve an element of judicial discretion, since the legislator can hardly have foreseen all the relevant circumstances of every case arising under the law. 'There are many situations', writes Dean Pound, 'where the course of judicial action is left to be determined wholly by the judge's individual sense of what is right.'[8] It would perhaps have been fairer to say that the good judge will be guided in such cases not so much by his own 'individual sense of what is right' as by the sense of right generally accepted by the community whose servant he is. But that some 'sense of what is right', whether individual or general, is a necessary ingredient of many judicial decisions, few will care to deny. The importance of the political presuppositions which inspire the Supreme Court of the United States in the interpretation of the Constitution, and the way in which, in the course of American history, these presuppositions have changed in response to changing social conditions, is well known.[9] The problem is, in its final analysis, the fundamental one of the relation of the rights of the individual to the needs of the community. Every national community has necessarily found a working solution of this problem. The international community has not yet done so. The controversy about the freedom of the seas shows that Great Britain would be unwilling to risk any interpretation of her maritime rights by an international court in the light of the supposed needs of the international community as a whole; and there are important matters on which every other Great Power would make similar reservations. The absence of an accepted view of the general good of the community as a whole overriding the particular good of any individual member of it, which we have already noted as the crucial problem of international morality, also stands in the way of the development of judicial settlement in its application to international disputes.

We find, therefore, in the problem of the justiciability of international disputes another illustration of the fact that law is a function of political society, is dependent for its development on the development of that society, and is conditioned by the political presuppositions which that society shares in common. It follows that the strengthening of international law, and the extension of the number and character of international disputes recognized as suitable for judicial settlement, is a political, not a legal, problem. There is no principle of law which enables one to decide that a given issue is suitable for treatment by legal methods. The decision is political; and its character is likely to be determined by the political development of the international community or of the political relations between the countries concerned. Similarly, there is no principle of law which enables one to decide whether a rule of law or a legal institution which has proved its value in a national community should be introduced by analogy into international law. The sole valid criterion is whether the present stage of political development of the international community is such as to justify the introduction of the rule or institution in question. In modern international relations, the machinery of judicial settlement has been developed far in advance of the political order in which alone it can effectively operate. Further progress towards the extension of the judicial settlement of international disputes can be made, not by perfecting an already too perfect machinery, but by developing political co-operation. The fact that the members of the British Commonwealth of Nations have hitherto steadfastly refused to set up any kind of permanent and obligatory procedure for the judicial settlement of disputes between one another should serve as a warning to those who are disposed to attach undue importance to the perfection of judicial machinery in international relations. It is a curious paradox that, by signing the Optional Clause of the Statute of the Permanent Court and by excluding from its operation inter-Commonwealth disputes, Great Britain and Dominions are bound in this respect towards many foreign countries by an obligation more far-reaching than they have assumed among themselves.

Projects of 'all-in arbitration'

Many thinkers of the period between the two wars went, however, far beyond mere plans for the modest and gradual extension of the scope of judicial procedure in international relations. It became a widely cherished ambition to provide, by a stroke of the pen, for the compulsory settlement of all international disputes by arbitration. Schemes for obligatory arbitration were mooted on many occasions prior to 1914, but failed to win acceptance. The Covenant of the League of Nations, while providing for the establishment of the Permanent Court and encouraging the submission of suitable disputes to arbitration or judicial settlement, gave little encouragement to the advocates of obligatory arbitration. In all disputes, it left the choice of the procedure to the discretion of the states concerned; and the political procedure of 'enquiry by the Council' always remained open. It was precisely this political aspect of the Covenant which became a target for the attacks of the utopian school. A widespread feeling grew up that the way to establish an international 'rule of law' and avoid future wars was for states to submit all international disputes of every kind to an international arbitral tribunal having power to decide them at its discretion on grounds either of strict law or of equity and common sense. Such was the vague conception summed up in the popular catchword of 'all-in arbitration'. This demand for 'all-in arbitration' was supposed to have been met by the Geneva Protocol and by the General Act. It was widely believed that, had the British Government not rejected the Protocol, or had the General Act been accepted without reservations by the principal Powers, a satisfactory procedure would have been in existence for the compulsory arbitration of all international disputes and an important cause of war removed.

But here we come upon an extraordinary confusion, or series of confusions, of thought which, throughout this period, enveloped and obscured the problem of the peaceful settlement of international disputes. When the League Covenant, by an amendment inserted after the establishment of the Permanent Court of International Justice, set 'judicial settlement' side by side with 'arbitration', 'arbitration' meant the verdict of a judge or a tribunal appointed *ad hoc*, and 'judicial settlement' the verdict of a regularly constituted court; and there is no reason to suppose that any other distinction was intended between them. But the misguided attempt to discover an objective distinction between justiciable and non-justiciable disputes led to an equally fallacious distinction between 'judicial settlement', meaning the settlement of 'justiciable' disputes in accordance with the letter of the law, and 'arbitration', meaning the settlement of 'non-justiciable' disputes, which were not covered by the letter of the law, on grounds of equity. This conception left its traces on the Geneva Protocol. According to the Assembly report on that instrument, 'the arbitrators need not necessarily be jurists', and if they obtain an advisory opinion on any point of law from the Permanent Court, that opinion is 'not legally binding on them'.[10] But the distinction between 'judicial settlement' and 'arbitration' was first fully developed in the General Act. Under this instrument, disputes 'with regard to which the parties are in conflict as to their respective rights' were to be referred to the Permanent Court for 'judicial settlement'. All other international disputes were to be referred for 'arbitration' to an arbitral tribunal. In the absence of any agreed stipulation to the contrary, the tribunal, in pronouncing its judgement was to apply the same rules of law as were applied by the Permanent Court. But 'in so far as there exists no such rule applicable to the dispute, the tribunal shall decide *ex aequo et bono*'. This reference to rules of law seems incomprehensible. If the dispute turned on legal rights, it would be submitted not to the arbitral tribunal, but to the Permanent Court. If it did not turn on legal rights, the dispute could not be solved by the application of legal rules. The conception that there is a class of international disputes which arise, so to speak, *in vacuo*, and are not affected by any existing legal rights or by any rule of international law, is a pure myth.

A more serious confusion is, however, in store. There is a perfectly valid distinction, familiar both in national and in international affairs, between 'legal' disputes, arising out of claims which purport to be based on existing legal rights, and 'political' disputes arising out of claims to alter existing legal rights. The difference turns, however, not on the nature of the dispute, but on the question whether the complainant seeks his remedy through legal or through political procedure. In the state, claims of the former kind are dealt with by the courts, claims of the latter kind by political action. The individual who fails to get his grievance remedied by a court may seek a remedy for the same grievance through legislation. Internationally, the distinction is less clear cut. No international court is recognized as competent to settle all 'legal' disputes, and there is no recognized machinery to settle all 'political' disputes. In these circumstances, states making claims against other states are not obliged to make it clear, and do not always make it clear, whether the claim is based on legal rights or is tantamount to a demand to alter those rights. But the distinction, though sometimes obscured in practice, is real enough. Both nationally and internationally, 'political' disputes are, generally speaking, more serious and more dangerous than 'legal' disputes. Revolutions and wars are less likely to arise from disputes about existing legal rights than from the desire to change those rights. The wise politician, and the wise student of politics, will devote a great deal of attention to political disputes.

When, therefore, it was officially claimed that the Geneva Protocol constituted 'a system for the pacific settlement of *all disputes* which might ever arise',[11] or that the General Act provided 'a comprehensive method of settling all international disputes of whatever character',[12] the conclusion might reasonably have been drawn, and was in fact drawn by many people, that provision had been made for the settlement by arbitration of political disputes, i.e. of disputes arising from claims to modify existing legal rights. Closer inspection did not, however, justify this conclusion. In an inconspicuous passage of the Assembly report on the Protocol, it was explained that the procedure did not apply to 'disputes which aim at revising treaties and international acts in force or which seek to jeopardize the existing territorial integrity of signatory states'. In fact, added the *rapporteur*, 'the impossibility of applying compulsory arbitration to such cases was so obvious that it was quite superfluous to make them the subject of a special provision'.[13] The General Act is less ingenuous. It purports to enforce compulsory arbitration for disputes which are not disputes about the 'respective rights' of the parties. It purports to authorize the arbitral tribunal to decide such disputes *ex aequo et bono*. But the authorization applies only 'in so far as there exists no [legal] rule applicable to the dispute'; and this qualification has the same effect as the reservation in the report on the Geneva Protocol. The essence of a political dispute is the demand that the relevant legal rule, though admittedly applicable, shall not be applied. When a dispute arises through the claim of a state that its existing frontiers, or existing treaty restrictions on its sovereignty, or existing obligations under a financial agreement, are intolerable, it is useless to refer it to an arbitral tribunal whose first duty is to apply the legal 'rule applicable to the dispute'. The legal right exists and is uncontested. The dispute arises from a demand to change it. Political disputes cannot be settled within the framework of the law by tribunals applying rules of law. The Geneva Protocol and General Act, though purporting to provide for the peaceful settlement of all international disputes, in fact left the most important and dangerous category of international disputes untouched.

No scheme of 'all-in arbitration' more inclusive than the make-believe of the Geneva Protocol and the General Act was officially propounded or considered. Some governments were prepared to accept arbitration for such disputes as did not endanger the existing political order – a limitation hardly less restrictive than the vital interests, independence and national honour of the older arbitration treaties. But no government was willing to entrust to an international court the power to modify its legal rights. Some theorists, however, were more ready than practical statesmen to brush this difficulty aside, and were quite prepared to entrust to a so-called arbitral tribunal the task not only of applying existing rights, but of creating new ones. A British organization called the New Commonwealth Society evolved an elaborate scheme for an arbitral tribunal which would 'determine, on the basis of equity and good conscience, political disputes, including those which have to do with the revision of treaties', thus establishing 'an indirect method of legislation in the affairs of nations' by an equity tribunal.[14] Such a scheme would appear to be the necessary corollary of Professor Lauterpacht's belief that international 'conflicts of interests are due ... to the imperfections of international legal organisation'.[15] International conflicts of interests will in future be resolved by a tribunal which will become the supreme organ of world government, exercising not merely the judicial function of interpreting the rights of states, but the legislative function of changing them. Thus will be realized another distinguished international lawyer's dream of 'an international legal community whose centre of gravity is in the administration of international justice'.[16]

These theories have one important merit. They recognize the fallacy, implicit in the Geneva Protocol and the General Act, that an international legal order based on the recognition, interpretation and enforcement of existing rights is an adequate provision for the peaceful settlement of international disputes. But in avoiding this fallacy, they fall into a still graver one. Perceiving that provision must be made for the modification of existing rights, they force this essentially political function into a legal mould and entrust its exercise to a tribunal. Unwilling to recognize the political basis of every legal system, they dissolve politics into law. In this quasi-judicial twilight, the judge becomes the legislator, political issues are settled by an impartial tribunal on grounds of equity and common sense, and the distinction between law and politics disappears.

The extreme difficulty of the international problem is no doubt responsible for the prescription of so heroic a remedy. But the fact that the problem is difficult scarcely justifies us in propounding a solution which nobody regards as either feasible or desirable in our far more highly organized national communities. The obligatory arbitration of international disputes of all kinds is, according to Professor Lauterpacht, '*a sine qua non* of the normal machinery for the preservation of peace'.[17] Yet obligatory arbitration of claims not based on legal right is rarely enforced in civilized states, and least of all in those which enjoy the longest record of domestic peace. It does not occur to us to attribute 'conflicts of interests' in our domestic politics to the imperfections of our legal organization, or to submit to a national arbitral tribunal, for impartial decision on grounds of equity and common sense, disputes about the necessity of conscription, the abolition of the means test, the legal status of trade unions, or the nationalization of mines. The difficulty is not that we could not find a group of impartial persons deeply imbued with the principles of equity and common sense, but that impartiality, equity and common sense are not the primary, or at any rate not the sole, qualities which we require in a decision of such issues. These are political issues, and are settled by procedure which allows for the intrusion of power, whether in the form of a majority vote, as in democracies, or of the will of a dictator or a party, as in authoritarian states. Neither in democracies nor in authoritarian states are such issues decided by an 'impartial' tribunal.

The inapplicability of judicial procedure to 'political' disputes

Why then is it necessary, not only in theory, for the sake of clear thinking, but also in practice, for the sake of good government, to preserve this distinction between the legal and political, between issues which we are willing to have settled by judicial procedure on grounds of existing legal rights, and issues which can only be settled by political procedure because they turn on a demand for the modification of existing legal rights?

The first answer is that judicial procedure differs fundamentally from political procedure in excluding the factor of power. When a dispute is submitted to a court, the presupposition is that any difference in power between the parties is irrelevant. The law recognizes no inequality other than inequality of legal right. In politics, the converse presupposition holds. Here power is an essential factor in every dispute. The settlement of a conflict of interest between British agriculturalists and British industrialists will depend, in part at any rate, on their respective voting strength and the respective 'pulls' which they can exercise on the government. The settlement of a conflict of interest between the United States and Nicaragua will depend, in greater part (for the ratio of power to other factors is higher in international than in national politics), on the relative strength of the two countries. Conflicts of interest can be dealt with only by an organ which takes the power factor into account. Nothing is gained, and the proper function of law is debased and discredited, if this political function is entrusted to a tribunal whose constitution and procedure are deliberately assimilated to those of a court of law. As Mr Bernard Shaw has remarked, the functions of judge and legislator are 'mutually exclusive': the former must ignore every interest, the latter take every interest into account.[18]

The second answer is equally fundamental. We have seen that even the strictly judicial procedure of a court sometimes entails political presuppositions, if only because the application of the law to the particular case is always liable to involve an element of judicial discretion, and this discretion, if it is not to be purely capricious, must draw its inspiration from those presuppositions. Where a tribunal is called upon to decide not on issues of legal right, but on claims to set aside legal rights in favour of equity or common sense, the necessity of clearly defined political presuppositions becomes all the more obvious. In such cases, judicial discretion instead of being limited to points left ambiguous by the law, has infinite scope; and the decisions of the tribunal, if they are not to be mere expressions of individual opinion, must be based on well-established assumptions shared by the community as a whole or by those who speak in its name. The existence of such assumptions in national communities sometimes makes possible the use of arbitration even in political issues; and the same possibility is not entirely excluded in the international sphere. But generally speaking, it is a fundamental obstacle to international arbitration *ex aequo et bono* that common assumptions of a far-reaching kind scarcely exist in the international community. To submit to an international tribunal, for decision on grounds not of law, but of equity and common sense, disputes concerning British interests in Egypt or the interests of the United States in the canal zone of Panama, or the future of Danzig, or the frontiers of Bulgaria, would have been impracticable, not only because the settlement of these problems involves issues of power, but also because there is no political agreement even of the vaguest kind as to what equity and common sense mean in relation to such questions. On the rare occasions on which international tribunals have been empowered by the parties to decide issues between them on grounds other than those of strict law, the tribunals have shown the greatest reluctance to avail themselves of the discretion accorded to them; not, as Professor Lauterpacht supposes, because 'law is more just than loose conceptions of justice and equity',[19] but because no responsible tribunal cares to commit itself on any important issue to an authoritative pronouncement as to what is 'equitable' or 'just' in international relations. An international tribunal, once it has left the comparatively solid ground of international law and legal rights, can find no foothold in any agreed conception of equity or common sense or the good of the community. It remains, in Professor Zimmern's words, 'an array of wigs and gowns vociferating in emptiness'.[20]

The crux, however, remains. Political issues, both nationally and internationally, are far more menacing than issues of legal right. The periodical, or rather the constant, revision of existing rights is one of the prime necessities of organized society; and to bring about revision in the international society by means other than war is the most vital problem of contemporary international politics. The first step has been to extricate ourselves from the blind alley of arbitration and judicial procedure, where no solution of this problem is to be found. Having taken this step, we are free to approach it by other, and perhaps more promising, avenues.

Notes

1. *Permanent Court of International Justice*, Series 2, No. 5, p. 27.
2. Disputes 'as to the interpretation of a treaty' are the first category of dispute recognized by the Covenant as 'generally suitable' for judicial settlement. It is noteworthy that the framers of the Covenant, who drew up this article, nevertheless rejected a proposal to insert in the Covenant a provision that disputes as to its own interpretation should be submitted to the Permanent Court (Miller, *The Drafting of the Covenant*, ii. pp. 349, 516). Behaviour in concrete cases is sometimes more significant than the enunciation *in vacuo* of abstract rules.
3. Lauterpacht, *The Function of Law in the International Community*, p. 369 and *passim*. It is a pity that Professor Lauterpacht, having brilliantly conducted his analysis up to the point where the unwillingness of states is recognized as the limiting factor in the justiciability of international disputes, should have been content to leave it there, treating this 'unwillingness', in true utopian fashion, as perverse and undeserving of the attention of an international lawyer.
4. The British Government, in its memorandum of 1928 on arbitration (*League of Nations: Official Journal*, pp. 694–704), criticized general arbitration treaties on the ground that, in the case of every country, 'obligations which it may be willing to accept towards one state it may not be willing to accept towards another'.
5. Quoted in Taracouzio, *The Soviet Union and International Law*, p. 296.
6. *Proceedings of the Hague Peace Conference* (Engl. transl.: Carnegie Endowment), *Conference of 1907*, ii. p. 316.
7. On the latter occasion, Professor Gilbert Murray, representing South Africa on the Assembly of the League of Nations, lamented that a judicial question (i.e. compensation for Italy) had been brought before a political organ and decided on political grounds (*League of Nations: Fourth Assembly*, pp. 139 *sqq.*) – an excellent example of the fallacy, so trenchantly exposed by Professor Lauterpacht, that certain issues are *ipso facto* judicial.

8. Roscoe Pound, *Law and Morals* (2nd ed.), p. 62.
9. Professor Laski remarked many years ago that 'the foreigner in the United States cannot but observe with the deepest wonder how eagerly possible nominations for a vacant position on the Supreme Court are canvassed' (Introduction to English translation of Duguit, *Law in the Modern State*, p. xxiii). The wonder has grown less deep since the political character of the Court has been better understood.
10. *League of Nations: Fifth Assembly*, First Committee, p. 486.
11. *League of Nations: Fifth Assembly*, p. 497.
12. *Memorandum on the General Act*, Cmd 3803, p. 4.
13. *League of Nations: Fifth Assembly*, p. 194.
14. Lord Davies, *Force*, pp. 73, 81.
15. Lauterpacht, *The Function of Law in the International Community*, p. 250.
16. Kelsen, *The Legal Process and International Order*, p. 30.
17. Lauterpacht, *The Function of Law in the International Community*, p. 438.
18. G. B. Shaw, *John Bull's Other Island*, Preface.
19. Lauterpacht, *The Function of Law in the International Community*, p. 252.
20. Zimmern, *The League of Nations and the Rule of Law*, p. 125. The words are applied to Taft's international arbitral court. They could be applied, still more appositely, to the equity tribunal advocated by the New Commonwealth Society.

CHAPTER THIRTEEN

Peaceful Change

Recognition of the need for political change has been a commonplace of thinkers of every period and every shade of opinion. 'A state without the means of some change', said Burke in a famous phrase, 'is without the means of its own conservation.'[1] In 1853, Marx wrote trenchantly on the Eastern question:

> Impotence ... expresses itself in a single proposition: the maintenance of the *status quo*. This general conviction that a state of things resulting from hazard and circumstances must be obstinately maintained is a proof of bankruptcy, a confession by the leading Powers of their complete incapacity to further the cause of progress and civilization.[2]

And Professor Gilbert Murray has put the same point in a different form:

> War does not always arise from mere wickedness or folly. It sometimes arises from mere growth and movement. Humanity will not stand still.[3]

It appears to follow from this view that the attempt to make a moral distinction between wars of 'aggression' and wars of 'defence' is misguided. If a change is necessary and desirable, the use or threatened use of force to maintain the *status quo* may be morally more culpable than the use or threatened use of force to alter it. Few people now believe that the action of the American colonists who attacked the *status quo* by force in 1776, or of the Irish who attacked the *status quo* by force between 1916 and 1920, was necessarily less moral than that of the British who defended it by force. The moral criterion must be not the 'aggressive' or 'defensive' character of the war, but the nature of the change which is being sought and resisted. 'Without rebellion, mankind would stagnate and injustice would be irremediable.'[4] Few serious thinkers maintain that it is always and unconditionally wrong to start a revolution; and it is equally difficult to believe that it is always and unconditionally wrong to start a war. Everyone will, however, agree that war and revolution are undesirable in themselves. The problem of 'peaceful change' is, in national politics, how to effect necessary and desirable changes without revolution and, in international politics, how to effect such changes without war.

Every effective demand for change, like every other effective political force, is compounded of power and morality; and the object of peaceful change can be expressed in terms neither of pure power nor of pure morality. It is rather unprofitable, except as an academic exercise, to enquire whether the purpose of any change should be to establish 'justice', by remedying 'just' grievances, or to maintain 'peace', by giving satisfaction to those forces which would otherwise be strong enough to make revolution or war. But it is dangerous to suppose that the two purposes are identical, and that no sacrifice of one or the other is required. Every solution of the problem of political change, whether national or international, must be based on a compromise between morality and power.

The role of power in political change

The necessary role of power in political change will be ignored only by the most superficial observers. Few 'revisionist' campaigns in history have been more firmly based on moral considerations than that of the Dreyfusards in France. Yet the protest against the condemnation of Dreyfus would never have been effective if it had not been taken up by powerful political organizations and used by them as a weapon against political opponents. The grievances of Albania and Nicaragua, whatever their moral basis, will never be effective unless they are endorsed, for interested reasons, by some Great Power or Powers. It is fair to attribute the growth of social legislation in the last hundred years to a growing realization of the just grievances of the working class. Yet these results would never have been achieved without the constant use, or threatened use, of force in the form of strikes and revolutions. 'It is true', remarks Mr John Strachey, 'that governments always tell us that they will never yield to force. All history tells us, however, that they never yield to anything else.'[5] 'Peaceful secession!' exclaimed Daniel Webster in 1849. 'Sir, your eyes and mine are never destined to see that miracle.'[6] 'The winning back of the lost territories', wrote Hitler in a famous passage of *Mein Kampf*, 'is not achieved through solemn invocations of the Lord God or through pious hopes in a League of Nations, but through armed force.'[7] Hitler might even have appealed to the respectable authority of Gladstone who, in the days when liberalism was still a political force, observed that 'if no considerations in a political crisis had been addressed to the people of this country except to remember to hate violence and love order and exercise patience, the liberties of this country would never have been obtained'.[8] It has been said that no ruling class ever abdicates of its own accord. Article 19 of the Covenant of the

League of Nations remains a lonely monument to the pathetic fallacy that international grievances will be recognized as just and voluntarily remedied on the strength of 'advice' unanimously tendered by a body representative of world public opinion.

While, however, the fundamental problem of political change – the compromise between power and morality – is identical in national and in international politics, the question of procedure is complicated by the unorganized character of the international community. Analogies drawn from procedures of change in the national sphere can only be applied with caution to the international field. We have seen that judicial procedures cannot be invoked, either nationally or internationally, for the solution of ultimate political problems. But the analogy of legislation seems at first sight more hopeful. The legislative process, unlike the judicial process, recognizes the role of power which is inherent in all political change (for the legislative authority is the supreme power of the state imposing its will on the whole community); and legislation, which a German writer has called 'legal revolution',[9] is the most obvious and regular way of bringing about political change within the state. 'What is peaceful change as an effective institution of international law or of international society?' asks Professor Lauterpacht, and answers: 'It is the acceptance by states of a legal duty to acquiesce in changes in the law decreed by a competent international organ.'[10]

It has already been noted that international law rests upon custom, and that there is at the present time no such thing as international legislation or an international legislature. The terms of Article 19 of the Covenant show how remote the principal states were in 1919 from 'acceptance of a legal duty to acquiesce in changes in the law decreed by a competent international organ'. Nor can this well be otherwise. Reflexion will show that the legislative process, like the judicial process, presupposes the existence of a political order. It is only by that combination of consent and coercion which underlies every political society that we can arrive at the establishment of a supreme organ, whether parliament or council of state or individual autocrat, whose fiat creates law binding on all members of the community. These conditions are not fulfilled in the international community. The Assembly of the League of Nations, whose decisions required unanimity, was a conference empowered to conclude international agreements, not a legislature which passed international legislation; for, as Mr Eden bluntly observed at one of its sessions, 'it would plainly be impracticable ... to give the Assembly power to impose changes against the wish of the parties concerned'.[11] The difficulty lies not in the lack of machinery for international legislation, but in the absence of an international political order sufficiently well integrated to make possible the establishment of a legislative authority whose decrees will be recognized as binding on states without their specific assent. If we accept Professor Lauterpacht's identification of peaceful change with international legislation, we can only conclude that, in his words, 'an international system of peaceful change ... runs the risk of being unreal unless it forms part and parcel of a comprehensive political organization of mankind'.[12] The condition of international legislation is the world super-state.

Need we, however, reconcile ourselves to the discouraging conclusion that any international system of peaceful change must await the coming of the super-state? The analogy of legislation may turn out to be not merely discouraging, but misleading. The present almost universal belief in the beneficence of legislation as a reforming instrument within the state is in the main a growth of the past fifty years. Down to the end of the nineteenth century, many intelligent people continued to regard the state as a necessary evil and legislation as a regrettable device not to be resorted to except in case of proven necessity.[13] Within the national community, the distinction familiar to nineteenth-century thought between 'society' and 'state' has lost much of its significance through the development of the social functions of the modern state. But in the international sphere, we are in the presence of a 'society' which has no corresponding 'state'; and we may therefore find some help in the conception, which would hardly have seemed paradoxical to any age but our own, of changes peacefully effected in the social structure without legislation or any other overt form of state intervention. Even to-day, it is easy to exaggerate the role of legislation; and it may still be true to say (as it would certainly have been true a hundred years ago) that the most important changes in the structure of society and in the balance of forces within it are effected without legislative action. It may be unnecessarily pessimistic to rush into the conclusion that the absence of an international legislature rules out any international procedure of peaceful change.

If, therefore, we are looking for the nearest analogy in the national community to the turbulent relations which render the problem of change acute in the international society, we may find it in the relations of those group-entities within the state whose conflicts have not been in the past, and still in large measure are not, settled by any legislative process. Of these group-entities, by far the most important, and by far the most instructive for our purpose, are those representing capital and labour respectively. Here we have the same recurrent conflict between 'haves' and 'have-nots', between 'satisfied' and 'dissatisfied'; the same reluctance, on the part of one or both sides, to accept the principle of 'all-in arbitration' for the settlement of their disputes; the same recognition of the inapplicability or inadequacy of the legislative process; the same appeals to 'law and order' by the satisfied group; and the same use, or threatened use, of violence by the dissatisfied in order to assert their claims. It is sometimes said that there can be no international procedure of peaceful change so long as states insist on being judges in their own cause. Here is a class of disputes in which both parties commonly insist on being judges in their own cause, and in which some progress at least has been made towards an orderly procedure of peaceful change.

Force has always been a crucial factor in relations between capital and labour. In the beginnings of the industrial revolution, every attempt at organized self-help on the part of the workers was rigorously repressed. This unqualified repression ended in Great Britain with the repeal of the Combination Acts in 1825, and continued in Russia as late as 1905. Between those two dates, the workers of every important industrial country secured recognition of their right to use the weapon of the organized strike. The strike not only proved itself an effective instrument for extracting concessions from employer to workers, but became a recognized symbol of the major weapon of force – revolution.[14] In recent times, the element of force has been once more eliminated from relations between employer and workers by the authoritarian governments of Soviet Russia,[15] Italy and Germany, through legislation prohibiting strikes and an executive strong and ruthless enough to enforce the prohibition. Democratic countries have from time to time prohibited strikes, though such prohibitions have nearly always been resisted by the workers, and have rarely proved enforceable over an extended period.[16] Theoretically, force might in the same way be eliminated from the settlement of international disputes by a powerful and authoritarian super-state. But this result, whether desirable or not, lies outside the scope of practical consideration; and we shall therefore find a better analogy to the international position if we consider those countries and those periods in which relations between capital and labour have not been dominated by the overwhelming power of the state. In the latter part of the nineteenth century and the first part of the twentieth the 'have-nots' of most countries steadily improved their position through a series of strikes and negotiations, and the 'haves', whether through a sense of justice, or through fear of revolution in the event of refusal, yielded ground rather than put the issue to the test of force. This process eventually produced on both sides a willingness to submit disputes to various forms of conciliation and arbitration, and ended by creating something like a regular system of 'peaceful change'. In many countries such a system has been in operation for many years with remarkable success, though the ultimate right to resort to the weapon of the strike is not abandoned. If we could apply this analogy to international relations, we might hope that, once the dissatisfied Powers had realized the possibility of remedying grievances by peaceful negotiations (preceded no doubt in the first instance by threats of force), some regular procedure of 'peaceful change' might gradually be established and win the confidence of the dissatisfied; and that, once such a system had been recognized, conciliation would come to be regarded as a matter of course, and the threat of force, while never formally abandoned, recede further and further into the background. Whether the analogy is in fact valid, or whether this hope is purely utopian, is a question which can hardly be settled except by the test of experience. But one may record with some confidence the view that this is the only line of advance which affords any prospect at all of the establishment of any international procedure, however imperfect, of peaceful change.

The implication of this procedure must, however, be clearly recognized. Few issues of social or political change of sufficient magnitude to involve the risk of revolution or war can be settled without detriment, or apparent detriment, to the interests of one of the parties. That the party at whose expense the change was to be effected would acquiesce in it without the existence of means of pressure to compel him to do so was one of the strange illusions of the ill-fated Article 19 of the Covenant; and this illusion may be discarded. Such self-abnegation could indeed hardly have been expected. The statesman, the trade union leader or the company director is a trustee for those whose interests he represents; and in order to justify extensive concessions at their expense, he must generally be in a position to plead that he has yielded to *force majeure*. When the change is effected by legislation, the compulsion is that of the state. But where the change is effected by the bargaining procedure, the *force majeure* can only be that of the stronger party. The employer who concedes the strikers' demands pleads inability to resist. The trade union leader who calls off an unsuccessful strike pleads that the union was too weak to continue. 'Yielding to threats of force', which is sometimes used as a term of reproach, is therefore a normal part of the process.

The parallel should not be pressed too far. The role of force, even in the most advanced democratic states, is indeed more constant and more conspicuous than most sentimental democrats care to admit. In so orderly a country as Great Britain, during the present century, force has been used or threatened for securing political ends by Ulstermen, Irish nationalists, female suffragists, communists, fascists and organized workers. But within the state there are checks on the too hasty resort to force. In the first place, the legislative process exists, and provides an alternative method of change; faith in the ballot-box has deterred the workers of many countries from revolutionary policies. Secondly, the state makes some show (often an imperfect one) of holding the balance impartially between the parties on the issue in dispute. Thanks to these checks, a certain moral discredit attaches in democratic countries, in the minds of all classes, to the open use or threat of force until other means have been tried of bringing about a change.

In international politics neither of these checks exists. The use or threatened use of force is therefore a normal and recognized method of bringing about important political change, and is regarded as morally discreditable mainly by those 'conservative' countries whose interests would suffer from change. The largest operation of 'peaceful change' in the nineteenth century was that performed by the Congress of Berlin, which revised the treaty imposed by Russia on Turkey at San Stefano. But this revision took place only under the tacit threat of a declaration of war against Russia by Great Britain and Austria-Hungary.[17] The Lausanne Treaty of 1923 was a revision, extorted by the use and threatened use of force, of the treaty signed with Turkey at Sèvres in 1920. It was denounced by Lloyd George as 'an abject, cowardly and infamous surrender'; and this opinion was widely held at the time.[18] The revision of Czecho-Slovakia's frontiers effected by the Munich Agreement of September 1938 was also the product of a threat of force; and here we have the explicit testimony of M. Benes that no alternative method was available. For five years earlier he had publicly stated 'that no country could be forced by anyone to revise its frontiers, and that anyone who attempted it in the case of Czecho-Slovakia would have to bring an army along'.[19] Another curious example may be added. When Poland annexed Vilna in 1920, Lithuania closed the frontier and severed all communications with Poland. It is doubtful whether this isolation conferred any advantage on Lithuania. But no Lithuanian statesman could have justified the reopening of the frontier and the consequent loss of face to his country if he had not been in a position to plead *force majeure*. In March 1938, Poland mobilized an army and presented an ultimatum to Lithuania. The frontier was at once reopened; and normal relations were established. An operation of peaceful change, generally recognized as salutary, could not be effected save under a threat of war. Normally, the threat of war, tacit or overt, seems a necessary condition of important political changes in the international sphere.[20]

This principle has not only been demonstrated in practice on many occasions, but received a large measure of theoretical recognition from the framers and interpreters of the Covenant of the League of Nations. The machinery of the League was brought into action by the danger of war. Article 11 related to 'any war or threat of war' and to 'any circumstance ... which threatens to disturb international peace'; and Article 19 purported to deal with treaties which have become inapplicable' (a phrase which has never been satisfactorily explained) and with 'international conditions whose continuance might endanger the peace of the world'. Moreover, the most effective article of the Covenant for promoting peaceful change, and the only one which was ever invoked for the purpose,[21] was not Article 19 but Article 15, under which recommendations could be made without the concurrence of the parties concerned, and might, in the event of war, be supported by sanctions. But the only condition which could bring this article into operation was a 'dispute likely to lead to a rupture'. The grievances of which the Covenant took cognizance were, broadly speaking, the grievances of states strong enough to create a danger of war. In 1932, when Finland brought before the Council a claim against Great Britain arising out of the commandeering of Finnish ships in the first world war, the British Government argued *inter alia* that there was no case to go to the Council, since the dispute was not in the least 'likely to lead to a rupture'. In the same year, the British Government brought before the Council under Article 15 a dispute with Iran arising out of the affairs of the Anglo-Iranian Oil Company. The essential difference was that Great Britain was strong enough to create the danger of a rupture, and Finland was not.[22] When Article 19 was invoked for the first time by Bolivia in 1921, it was cogently argued that, since the conditions of which Bolivia complained had existed for a long period without endangering peace, there was no case for bringing them before the League. In other words, it would have been necessary, in order to set the procedure of peaceful change in motion, that Bolivia should be strong enough to threaten war against Chile. The doctrine of the Covenant thus confirmed the lesson of experience that peaceful change could not be effected on any important scale in international politics in the absence of a threat, or potential threat, of war.

We may sum up the conclusions so far reached. The judicial process is unfitted to solve the problem of peaceful change in national, and *a fortiori* in international, politics; for, treating the parties to a dispute as equal, it fails to recognize the element of power which is a necessary factor in every demand for change. The legislative process, though recognizing the role of power and well adapted to meet many demands for change in national politics, is inapplicable to international demands for change, since it pre-supposes the existence of a legislative authority whose decrees are binding on all members of the community without their specific assent. There remains the bargaining process, which is applied to some demands for change within the state and is alone applicable to demands for international change, since states (like trade unions or employers' federations) insist on the ultimate right to accept or reject any solution offered. But whereas under the legislative process change is enforced by the power of the state, change under the bargaining process can be enforced only by the power of the complainant. Power, used, threatened or silently held in reserve, is an essential factor in international change; and change will, generally speaking, be effected only in the interests of those by whom, or on whose behalf, power can be invoked. 'Yielding to threats of force' is a normal part of the process of peaceful change.

This is one side of the picture; and since it is the side which is ignored in most modern writing about international politics, it has been deliberately emphasized here. Nor should we underrate the value of peaceful change even considered solely from this point of view. If relations between employers and workers are such that the former cannot resist, or the latter cannot sustain, a demand for an increase in wages and a reduction in hours, it is preferable (quite apart from any question of the justice or injustice of the demand) that it should be conceded or rejected as the result of peaceful negotiations rather than as the result of a long and embittered strike which half ruins both employers and workers. If the relations of power between the leading European countries in 1877 made it inevitable that Bulgaria should be deprived of much of the territory allocated to her by the Treaty of San Stefano, then it was preferable that this result should be brought about by discussions round a table in Berlin rather than by a war between Great Britain and Austria-Hungary on the one side and Russia on the other. If we consider peaceful change merely as a more or less mechanical device, replacing the alternative device of war, for readjusting the distribution of territory and of other desirable things to changes in the equilibrium of political forces, it performs a function whose utility it would be hypocritical to deny. Many changes made in national communities, whether by legislation or otherwise, and recognized as salutary, have no other basis than this.

The role of morality in political change

Nevertheless, it is clear that there is another aspect of peaceful change which occupies men's thoughts, and that it is no more possible to discuss peaceful change than to discuss any other kind of political procedure in terms of power alone. When a contested demand for change is made, the question which immediately exercises the minds of most people is whether it is just. It is true that our view of its justice is likely to be coloured, and may be wholly determined, by our own interest. It is true that, if our interest is not strongly engaged, we shall be tempted to discover reasons for regarding as just a solution which seems inevitable, or which could only be avoided by a great effort on our part. It is also true that, here as in every other political issue, power plays a part in determining our moral outlook, so that we shall be disposed, other things being equal, to regard a solution desired by the strong or the many as juster than a solution desired by the weak or the few. But when all these allowances have been made, the view taken of the morality of the transaction – a view not wholly determined by considerations of power – will influence the attitude of the mass of people affected by it. 'If orderly government is to command general assent,' writes Mr Bertrand Russell, 'some way must be found of persuading a majority of mankind to agree upon some doctrine other than that of Thrasymachus';[23] and if an orderly procedure of peaceful change is ever to be established in international relations, some way must be found of basing its operation not on power alone, but on that uneasy compromise between power and morality which is the foundation of all political life. The establishment of a procedure of peaceful negotiation in disputes between capital and labour presupposes, not merely an acute perception on both sides of the strength and weakness of their respective positions at any given time, but also a certain measure of common feeling as to what is just and reasonable in their mutual relations, a spirit of give and take and even of potential self-sacrifice, so that a basis, however imperfect, exists for discussing demands on grounds of justice recognized by both. It is the embryonic character of this common feeling between nations, not the lack of a world legislature, and not the insistence of states on being judges in their own cause, which is the real obstacle in the way of an international procedure of peaceful change.

How far is this common feeling operative in relation to demands for international change? Clearly in some degree. Two concrete cases of demands for change may be selected for analysis, one from the quasi-international, the other from the international sphere.

In the nineteenth century, the demand for home rule for Ireland found among a large number of people in Great Britain a support based not on considerations of power, but on common recognition as a canon of international morality of the right of 'oppressed nationalities' to self-determination and on a certain readiness to sacrifice self-interest to it. The stock of common feeling between Great Britain and Ireland was considerably greater than that commonly existing between two foreign countries. Nevertheless, the demand for change did not become effective until, owing to the diversion of British military strength elsewhere, force could be placed behind it. If the compromise eventually arrived at in 1921 could have been achieved in 1916, it would have been a true example of peaceful change achieved, like most international examples of peaceful change, under threat of war. But even in 1921, the settlement could not have been reached, and above all could not have been lasting, solely on a basis of power. The Anglo-Irish Treaty was a flagrant case of 'yielding to threats of force': it was concluded with the authors of a successful rebellion. But it had its necessary moral foundation in the acceptance of a common standard of what was just and reasonable in mutual relations between the two countries, and in the readiness of both (and particularly of the stronger) to make sacrifices in the interest of conciliation; and this made a striking success of an agreement about which the gloomiest prognostications were current at the time of its conclusion.

The second example is the failure to achieve a peaceful settlement with Germany in the period between the two world wars. The mass of political opinion in Great Britain and Germany (and in most other countries) agreed for many years that a criterion of justice and injustice could properly be applied to the Versailles Treaty; and there was a surprisingly considerable, though far from complete, consensus of opinion about the parts of it which were just and unjust respectively. Unfortunately, Germany was almost wholly deficient for fifteen years after 1918 in that power which is, as we have seen, a necessary motive force in political change; and this deficiency prevented effect being given, except on a minor scale, to the widespread consensus of opinion that parts of the Versailles Treaty ought to be modified. By the time Germany regained her power, she had adopted a completely cynical attitude about the role of morality in international politics. Even though she continued to base her claims on grounds of justice, she expressed them more and more clearly in terms of naked force; and this reacted on the opinion of the *status quo* countries, which became more and more inclined to forget earlier admissions of the injustices of the Versailles Treaty and to consider the issue as exclusively one of power.* The easy acquiescence of the *status quo* Powers in such actions as the denunciation of the military clauses, the reoccupation of the Rhineland or the annexation of Austria was due, not wholly to the fact that it was the line of least resistance, but in part also to a consensus of opinion that these changes were in themselves reasonable and just.[24] Yet they were greeted in each case by official censures and remonstrances which inevitably created the impression that the remonstrating Powers acquiesced merely because they were unable or unwilling to make the effort to resist. Successive removals of long recognized injustices of the Versailles Treaty became a cause not of reconciliation, but of further estrangement, between Germany and the Versailles Powers, and destroyed instead of increasing the limited stock of common feeling which had formerly existed.

It is beyond the scope of the present book to discuss the present or future foreign policy of Great Britain or of any other state. But the defence of the *status quo* is not a policy which can be lastingly successful. It will end in war as surely as rigid conservatism will end in revolution. 'Resistance to aggression', however necessary as a momentary device of national policy, is no solution; for readiness to fight to prevent change is

*This reaction was, of course, intensified by Nazi Germany's domestic policy.

just as unmoral as readiness to fight to enforce it. To establish methods of peaceful change is therefore the fundamental problem of international morality and of international politics. We can discard as purely utopian and muddle-headed plans for a procedure of peaceful change dictated by a world legislature or a world court. We can describe as utopian in the right sense (i.e. performing the proper function of a utopia in proclaiming an ideal to be aimed at, though not wholly attainable) the desire to eliminate the element of power and to base the bargaining process of peaceful change on a common feeling of what is just and reasonable. But we shall also keep in mind the realist view of peaceful change as an adjustment to the changed relations of power; and since the party which is able to bring most power to bear normally emerges successful from operations of peaceful change, we shall do our best to make ourselves as powerful as we can. In practice, we know that peaceful change can only be achieved through a compromise between the utopian conception of a common feeling of right and the realist conception of a mechanical adjustment to a changed equilibrium of forces. That is why a successful foreign policy must oscillate between the apparently opposite poles of force and appeasement.

Notes

1. Burke, *Reflections on the Revolution in France* (Everyman ed.), p. 19.
2. Marx and Engels, *Works* (Russian ed.), ix, p. 372.
3. Gilbert Murray, *The League of Nations and the Democratic Idea*, p. 16.
4. B. Russell, *Power*, p. 263.
5. J. Strachey, *The Menace of Fascism*, p. 228.
6. Quoted in J. Truslow Adams, *The Epic of America*, p. 239.
7. Hitler, *Mein Kampf*, p. 708.
8. Quoted in E. Pethick-Lawrence, *My Part in a Changing World*, p. 269.
9. Berber, *Sicherheit und Gerechtigkeit*, p. 9.
10. *Peaceful Change*, ed. C. A. W. Manning, p. 141.
11. *League of Nations: Seventeenth Assembly*, p. 46.
12. *Peaceful Change*, ed. C. A. W. Manning, p. 164.
13. Professor Lauterpacht's remark that 'the circle of interests directly regulated by law expands with the growth of civilization' (*The Function of Law in the International Community*, p. 392) is a truism to-day, but would have seemed a paradox to many nineteenth-century thinkers.
14. This is the significance of the 'one-day strike', which was popular in certain countries and which, though useless in itself, was designed to demonstrate that the workers were strong enough to break the power of the state. The success or failure of the one-day strike was thus a test of power, and its result enabled both sides to draw the appropriate conclusion without resorting to extreme measures.
15. The position is not substantially affected by the fact that in Soviet Russia the employer is normally a state trust or institution.

16. In Great Britain, strikes in munitions factories were prohibited during the first world war by the Munitions of War Acts. But though strikes occurred, the law was rarely if ever enforced, and it came to an end with the War. Under the Trade Disputes Act of 1927, political strikes were declared illegal, but no such case appears to have arisen since the act was passed. The situation in other countries is summarized in a pamphlet published by the American League for Industrial Democracy (*Shall Strikes be Outlawed?* by Joel Seidman), which concludes that 'labour feels that its right to strike is its surest guarantee of fair treatment' and that 'along the path of voluntary collective bargaining lies the greatest hope of satisfactory industrial relations'.

17. A writer who has surveyed the history of peaceful change down to 1914 records the conclusion that 'it is always wisest to face Europe with a *fait accompli*' (Cruttwell, *History of Peaceful Change*, p. 3).

18. D. Lloyd George, *The Truth About the Peace Treaties*, ii. p. 1351.

19. *The Times*, 26 April 1933, quoted by Professor Manning in *Politica*, December 1938, p. 363.

20. Those who assert that change effected under a threat of armed force is not 'peaceful change' are, of course, at liberty to define their terms how they please. But it should be noted that a definition thus restricted would equally exclude changes effected by a legislative or judicial process, if these required enforcement. If Czecho-Slovak territory had been transferred to Germany in September 1938 by a decision of the League Assembly or of an equity tribunal, enforced by mobilizing the armies of the League or an international police force, the change would not for that reason have had any better title to the epithet 'peaceful'. Armed force would have been used in precisely the same way.

21. The Special Assembly, dealing with the Manchurian dispute under Article 15 of the Covenant, endorsed the recommendations of the Lytton Commission for substantial modifications of the *status quo* in Manchuria. It need hardly be added that Japanese military action was the force which prompted these recommendations, which proved, however, insufficient to satisfy Japan.

22. In the Finnish question, M. Madariaga expressed the view that 'it was extremely dangerous for the Council, the Assembly, and the League of Nations to establish the doctrine that irascible parties would be listened to, and calm parties would not, because in the latter case there would be no question of a rupture' (*League of Nations: Official Journal*, November 1934, p. 1458). The defect of the Finnish case was, however, not so much that Finland was calm as that she was weak.

23. B. Russell, *Power*, p. 100.

24. In Great Britain, a perusal of the British press for 7 and 8 March 1936 will show how widely the reoccupation of the Rhineland was not merely tolerated, but welcomed. Subsequently, the tone of the press became less favourable, being manifestly influenced by the more critical official attitude.

Conclusion

CHAPTER FOURTEEN

The Prospects of a New International Order

The end of the old order

Periods of crisis have been common in history. The characteristic feature of the crisis of the twenty years between 1919 and 1939 was the abrupt descent from the visionary hopes of the first decade to the grim despair of the second, from a utopia which took little account of the reality to a reality from which every element of utopia was rigorously excluded. The mirage of the nineteen-twenties was, as we now know, the belated reflexion of a century past beyond recall – the golden age of continuously expanding territories and markets, of a world policed by the self-assured and not too onerous British hegemony, of a coherent 'Western' civilization whose conflicts could be harmonized by a progressive extension of the area of common development and exploitation, of the easy assumptions that what was good for one was good for all and that what was economically right could not be morally wrong. The reality which had once given content to this utopia was already in decay before the nineteenth century had reached its end. The utopia of 1919 was hollow and without substance. It was without influence on the future because it no longer had any roots in the present.

The first and most obvious tragedy of this utopia was its ignominious collapse, and the despair which this collapse brought with it. 'The European masses realized for the first time', said a writer before the second world war, 'that existence in this society is governed not by rational and sensible, but by blind, irrational and demonic forces.'[1] It was no longer possible to rationalize international relations by pretending that what was good for Great Britain was also good for Yugoslavia and what was good for Germany was also good for Poland, so that international conflicts were merely the transient products of avoidable misunderstanding or curable ill-will. For more than a hundred years, the reality of conflict had been spirited out of sight by the political thinkers of Western civilization. The men of the nineteen-thirties returned shocked and bewildered to the world of nature. The brutalities which, in the eighteenth and nineteenth centuries, were confined to dealings between civilized and uncivilized peoples were turned by civilized peoples against one another. The relation of totalitarianism to the crisis was clearly one not of cause, but of effect. Totalitarianism was not the disease, but one of the symptoms. Wherever the crisis raged, traces of this symptom could be found.

The second tragedy of the collapse of utopia, which proceeded from the first and further intensified it, was of a subtler kind. In the latter half of the nineteenth century, when the harmony of interests was already threatened by conflicts of increasing gravity, the rationality of the world was saved by a good stiff dose of Darwinism. The reality of conflict was admitted. But since conflict ended in the victory of the stronger, and the victory of the stronger was a condition of progress, honour was saved at the expense of the unfit. After 1919 only Fascists and Nazis clung openly to this outmoded device for rationalizing and moralizing international relations. But the Western countries resorted to an equally dubious and disastrous expedient. Smitten by the bankruptcy of the harmony of interests, and shocked by its Darwinian deviation, they attempted to build up a new international morality on the foundation, not of the right of the stronger, but of the right of those in possession. Like all utopias which are institutionalized, this utopia became the tool of vested interests and was perverted into a bulwark of the *status quo*. It is a moot point whether the politicians and publicists of the satisfied Powers, who attempted to identify international morality with security, law and order and other time-honoured slogans of privileged groups, do not bear their share of responsibility for the disaster as well as the politicians and publicists of the dissatisfied Powers, who brutally denied the validity of an international morality so constituted. Both these attempts to moralize international relations necessarily failed. We can accept neither the Darwinian doctrine, which identifies the good of the whole with the good of the fittest and contemplates without repugnance the elimination of the unfit, nor the doctrine of a natural harmony of interests which has lost such foundation in reality as it once had, and which inevitably becomes a cloak for the vested interests of the privileged. Both these doctrines have become untenable as the basis of international morality. Their breakdown has left us with no ready solution of the problem of reconciling the good of the nation with the good of the world community; and international morality is in the melting-pot.

In what direction can we look for a revival of international morality? It is, of course, possible that no such revival is in prospect and that the world is descending into one of those historical periods of retrogression and chaos in which the existing mould of society is riven asunder and from which new and familiar forms eventually emerge. If so, the experience is unlikely to be either brief or painless. Those who believe in world revolution as a short cut to utopia are singularly blind to the lessons of history; and the number of those who hold this belief appears to have diminished in recent years. There is no more reason to assume that the path lies through world revolution than to take refuge in blank despair. Our task is to explore the ruins of our international order and discover on what fresh foundations we may hope to rebuild it; and like other political problems, this problem must be considered from the standpoint both of power and of morality.

Will the nation survive as the unit of power?

Before considering the role of power in any new international order, we must first ask what will be the unit of power. The current form of international politics is due to the fact that the effective units are nation-states. The form of the future international order is closely bound up with the future of the group unit.

The French Revolution, which inaugurated the period of history now drawing to its close, raised the issue of the rights of man. Its demand for equality was a demand for equality between individuals. In the nineteenth century, this demand was transformed into a demand for equality between social groups. Marx was right in perceiving that the individual in isolation could not be the effective unit in the struggle for human rights and human equality. But he was wrong in supposing that the ultimate unit was the social class, and in discounting the cohesive and comprehensive qualities of the national unit. The great European figures of the later nineteenth century were Disraeli and Bismarck, who strove to weld together the 'two nations' into one through the agencies of the social service state, popular education and imperialism, refuted the taunt that 'the worker has no country', and paved the way for 'national labour', 'national socialism' and even 'national communism'. Before 1914 the demand for equality was already beginning in Western Europe to pass over from the issue of equality between classes to that of equality between nations. Italian writers had described Italy as a 'proletarian' nation, using the term in the sense of 'under-privileged'. Germany demanded equality in the form of her place in the sun, which must, as Bernhardi said, be 'fought for and won against a superior force of hostile interests and powers'.[2] In France, socialist and ex-socialist ministers appealed for industrial peace in the interests of national unity. Imperceptibly the struggle between classes was coming to seem, even to the workers themselves, less important than the struggle between nations. And the struggle for equality became, in accordance with the ordinary laws of political power, indistinguishable from the struggle for predominance.

This then is the basic reason for the overwhelming importance of international politics after 1919. The conflict between privileged and unprivileged, between the champions of an existing order and the revolutionaries, which was fought out in the nineteenth century within the national communities of Western Europe, was transferred by the twentieth century to the international community. The nation became, more than ever before, the supreme unit round which centre human demands for equality and human ambitions for predominance. Everywhere in Europe, national governments and one-party states made their appearance; and where party issues survived, they were thought of as something outmoded and deplorable – a blot on national unity which cried out to be erased. The inequality which threatened a world upheaval was not inequality between individuals, nor inequality between classes, but inequality between nations. 'Just as inequality of wealth and opportunity between the classes often led to revolutions,' said Mussolini, 'so similar inequality between the nations is calculated, if not peaceably adjusted, to lead to explosions of a much graver character.'[3] The new harmony which was required was not (as the philosophers of *laissez-faire* assumed) a harmony between individuals, and not (as Marx assumed when he denied the possibility of its realization) a harmony between classes, but a harmony between nations. To-day we need not make the mistake, which Marx made about the social class, of treating the nation as the ultimate group unit of human society. We need not pause to argue whether it is the best or the worst kind of unit to serve as the focus of political power. But we are bound to ask ourselves whether, and if so by what, it is likely to be superseded. Speculation on this subject falls naturally into two questions:

(a)   Are the largest and most comprehensive units of political power in the world necessarily of a territorial character?

(b)   If so, will they continue to take approximately the form of the contemporary nation-state?

The question whether the largest and most comprehensive power units must necessarily be territorial cannot receive a dogmatic answer applicable to all periods of history. At present, such units have a distinctively territorial form. It is easy to read past history as a gradual development leading up, with occasional relapses, to this consummation; and political power is probably never, even in the most primitive societies, entirely divorced from the possession of territory. Yet in many periods of history, of which the mediaeval is the most recent, power has been based ostensibly – and in part, really – on grounds other than those of territorial sovereignty. It was acceptance of the principle *cuius regio eius religio* which substituted the unit based on domicile for the unit based on religious allegiance, and thereby laid the foundation of the modern nation-state. In no previous period of modern history have frontiers been so rigidly demarcated, or their character as barriers so ruthlessly enforced, as to-day; and in no period, as we have already seen, has it been apparently so impossible to organize and maintain any international form of power. Modern technique, military and economic, seems to have indissolubly welded together power and territory. It is difficult for contemporary man even to imagine a world in which political power would be organized on a basis not of territory, but of race, creed or class. Yet the enduring appeal of ideologies which transcend the limits of existing political units cannot be ignored. Few things are permanent in history; and it would be rash to assume that the territorial unit of power is one of them. Its abandonment in favour of some other form of organized group power would, however, be so revolutionary that little that holds true of international politics in the present period would apply to the new dispensation. International relations would be supplanted by a new set of group relationships.

The question whether the territorial units of the future are likely to retain approximately their present form is one of more immediate practical importance. The problem of the optimum size of units – whether units of industrial or agricultural production or units of political and economic power – is one of the most puzzling and important of the present time; and the near future may well see striking developments. In the field of political power, two contrary tendencies may be observed.

In one direction, there is a clearly marked trend towards integration and the formation of ever larger political and economic units. This trend set in in the latter part of the nineteenth century, and appears to have been closely connected with the growth of large-scale capitalism and industrialism, as well as with the improvement of means of communication and of the technical instruments of power. The first world war threw this development into conspicuous relief.

Sovereignty, that is freedom to make decisions of wide historical importance [wrote Naumann in his famous book published in 1915], is now concentrated at a very few places on the globe. The day is still distant when there shall be 'one fold, one shepherd', but the days are past when shepherds without number, lesser or greater, drove their flocks unrestrained over the pastures of Europe. The spirit of large-scale industry and of super-national organization has seized politics . . . This is in conformity with centralized military technique.[4]

The interlude of 1918, when nationalism momentarily resumed its disintegrating role, proved – at any rate in Europe – a dangerous fiasco.

The multiplication of economic units added disastrously to the problems of the postwar period. Naumann with his *Mittel-Europa* proved a surer prophet than Woodrow Wilson with his principle of self-determination. The victors of 1918 'lost the peace' in Central Europe because they continued to pursue a principle of political and economic disintegration in an age which called for larger and larger units. The process of concentration still continued. The more autarky is regarded as the goal, the larger the units must become. The United States strengthened their hold over the American Continents. Great Britain created a 'sterling *bloc*' and laid the foundations of a closed economic system. Germany reconstituted *Mittel-Europa* and pressed forward into the Balkans. Soviet Russia developed its vast territories into a compact unit of industrial and agricultural production. Japan attempted the creation of a new unit of 'Eastern Asia' under Japanese domination. Such was the trend towards the concentration of political and economic power in the hands of six or seven highly organized units, round which lesser satellite units revolved without any appreciable independent motion of their own. On the other hand there is some evidence that, while technical, industrial and economic development within the last hundred years has dictated a progressive increase in the size of the effective political unit, there may be a size which cannot be exceeded without provoking a recrudescence of disintegrating tendencies. If any such law is at work, it is impossible to formulate it with any precision; and prolonged investigation would be necessary to throw any light on the conditions which govern the size of political and economic units. The issue is, however, perhaps likely to be more decisive than any other for the course of world history in the next few generations.

One prediction may be made with some confidence. The concept of sovereignty is likely to become in the future even more blurred and indistinct than it is at present. The term was invented after the break-up of the mediaeval system to describe the independent character of the authority claimed and exercised by states which no longer recognized even the formal overlordship of the Empire. It was never more than a convenient label; and when distinctions began to be made between political, legal and economic sovereignty or between internal and external sovereignty, it was clear that the label had ceased to perform its proper function as a distinguishing mark for a single category of phenomena. Discussion of such questions as whether the British Dominions are 'sovereign' Powers, or in whom 'sovereignty' of the mandated territories is vested, reveals the growing confusion. Such discussions are either legal arguments on the question what powers the authorities in those areas are constitutionally entitled to exercise (in which case the use of the term 'sovereignty' gives little help), or else arguments of pure form on the question whether it is convenient to use the label 'sovereignty' to describe situations which diverge to a greater or less extent from a common pattern. The concept of sovereignty becomes definitely misleading when, for instance, in a computation of the value of British colonial trade or British colonial investment, Egypt and Iraq are excluded on the ground that they are sovereign states. It is unlikely that the future units of power will take much account of formal sovereignty. There is no reason why each unit should not consist of groups of several formally sovereign states so long as the effective (but not necessarily the nominal) authority is exercised from a single centre. The effective group unit of the future will in all probability not be the unit formally recognized as such by international law. Any project of an international order which takes these formal units as its basis seems likely to prove unreal.

It may be well to add at this point that group units in some form will certainly survive as repositories of political power, whatever form these units may take. Nationalism was one of the forces by which the seemingly irreconcilable clash of interest between classes within the national community was reconciled. There is no corresponding force which can be invoked to reconcile the now seemingly irreconcilable clash of interest between nations. It is profitless to imagine a hypothetical world in which men no longer organize themselves in groups for purposes of conflict; and the conflict cannot once more be transferred to a wider and more comprehensive field. As has often been observed, the international community cannot be organized against Mars. This is merely another aspect of the dilemma with which the collapse of the spacious conditions of nineteenth-century civilization has confronted us. It seems no longer possible to create an apparent harmony of interests at the expense of somebody else. The conflict can no longer be spirited away.

Power in the new international order

Power is a necessary ingredient of every political order. Historically, every approach in the past to a world society has been the product of the ascendancy of a single Power. In the nineteenth century the British fleet not only guaranteed immunity from major wars, but policed the high seas and offered equal security to all; the London money market established a single currency standard for virtually the whole world; British commerce secured – it is true, in an imperfect and attenuated form – a widespread acceptance of the principle of free trade; and English became the *lingua franca* of four continents. These conditions, which were at once the product and the guarantee of British supremacy, created the illusion – and to some extent the reality – of a world society possessing interests and sympathies in common. The working hypothesis of an international order was created by a superior power. The hypothesis has been destroyed by the decline, relative or absolute, of that power. The British fleet is no longer strong enough to prevent war; the London market can enforce a single currency standard only over a limited area; free trade has wholly broken down; and if the English language retains, and has increased, its ascendancy, this is due to the fact that it is shared by Great Britain with other important countries. By what power can the international order be restored?

This question is likely to be answered by different nations in different ways. Most contemporary Englishmen are aware that the conditions which secured the overwhelming ascendancy of Great Britain in the nineteenth century no longer exist. But they sometimes console themselves with the dream that British supremacy, instead of passing altogether away, will be transmuted into the higher and more effective form of an ascendancy of the English-speaking peoples. The *pax Britannica* will be put into commission and become a *pax Anglo-Saxonica*, under which the British Dominions, standing halfway between the mother country and the United States, will be cunningly woven into a fabric of Anglo-American co-operation. This romantic idea goes back to the last years of the nineteenth century when Great Britain was already conscious of the growing burden of world supremacy, and when Cecil Rhodes had one of the first recorded visions of world empire based on an Anglo-American partnership. Oddly enough, it was an American Ambassador in London who, just before the war, gave the idea its most concrete expression. In 1913, Walter Hines Page proposed that President Wilson should visit London and conclude an Anglo-American alliance. 'I think', he added, 'the world would take notice to whom it belongs and – be quiet.'[5] The Washington Naval Treaty of 1922 was a more or less conscious bid by Great Britain for an equal partnership with the United States in the management of the world. The hope was reiterated again and again, with the reserves and the caution dictated by American susceptibilities, by British statesmen between the two world wars.

> I have always believed [said Lord Baldwin at the Albert Hall in May 1935] that the greatest security against war in any part of the world whatever, in Europe, in the East, anywhere, would be the close collaboration of the British Empire with the United States of America. The combined powers of the navies, the potential manpower, the immediate economic power of the combined blockade, and a refusal to trade or lend money would be a sanction that no power on earth however strong dare face. It may be a hundred years before that desirable end may be attained; it may never come to pass. But

sometimes we may have our dreams. I look forward to the future, and I see that union of forces for peace and justice in the world, and I cannot but think, even if men cannot advocate it openly yet, that some day and some time those who follow us may see it and know that the peace of the world is guaranteed by those who speak our tongue.[6]

The enormous growth of interest in Great Britain in everything relating to the United States shows what deep roots this ambition has struck in British hearts.

On the other side of the Atlantic, the picture necessarily looks rather different. Instead of an old firm, anxious to renew its strength by taking young blood into partnership, we have here a young and untried nation, reliant on its own strength, but still uncertain how far that strength will carry it. The United States did not, until the turn of the century, stake out their claim for recognition as a Great Power. But it was not long before leading Americans were beginning to see visions of world supremacy.

My dream [said Woodrow Wilson in a speech on Independence Day, 1914] is that as the years go by and the world knows more and more of America, it ... will turn to America for those moral inspirations which lie at the basis of all freedom, ... and that America will come into the full light of day when all shall know that she puts human rights above all other rights, and that her flag is the flag not only of America, but of humanity.[7]

The dream proved prophetic. In 1918 world leadership was offered, by almost unanimous consent, to the United States. The fact that it was then declined does not prove that it may not be grasped at some future time. If historical precedents count for anything, a *pax Americana* imposed on a divided and weakened Europe would be an easier contingency to realize than a *pax Anglo-Saxonica* based on an equal partnership of English-speaking peoples. But we are here in the realm of speculation, where the serious student cannot do more than canvass guesses and possibilities.

The necessary drawback about all conceptions of a world order depending on the ascendancy of a superior Power is that they ultimately involve recognition of the right of the strongest to assume world leadership. The *pax Romana* was the product of Roman imperialism, the *pax Britannica* of British imperialism. The 'good neighbour' policy of the United States in Latin America is not the antithesis, but the continuation and consequence of 'Yankee imperialism'; for it is only the strongest who can both maintain their supremacy and remain 'good neighbours'. There is no theoretical reason to refuse to other nations the right to aspire to world leadership.

Whoever really desires in his heart the victory of the pacifist conception of the world [writes Hitler in *Mein Kampf*] must devote himself by every means to the conquest of the world by the Germans. ... The pacifist, humanitarian idea will perhaps be excellent when the man superior to all others shall first have so conquered and subjugated the world that he becomes its sole master.[8]

The policy of Japan, as the Chinese delegate remarked at an Assembly of the League of Nations, was to establish a *pax Japonica* in the Far East.[9] The Englishman or the American is entitled to resist such ambitions. But he cannot resist them on universal grounds which will appeal to the German or the Japanese. The conception of a *pax Germanica* or a *pax Japonica*, i.e. of a world order dominated by Germany or Japan, was *a priori* no more absurd and presumptuous than the conception of a *pax Britannica* would have seemed in the reign of Elizabeth or of a *pax Americana* in the days of Washington and Madison. The only reason why it would seem absurd for Nicaragua or Lithuania to aspire to world leadership is that, according to any reasonable prognostication, these countries will never be strong enough to have the slightest hope of attaining such an ambition. To attempt to ignore power as a decisive factor in every political situation is purely utopian. It is scarcely less utopian to imagine an international order built on a coalition of states, each striving to defend and assert its own interests. The new international order can be built only on a unit of power sufficiently coherent and sufficiently strong to maintain its ascendancy without being itself compelled to take sides in the rivalries of lesser units. Whatever moral issues may be involved, there is an issue of power which cannot be expressed in terms of morality.

Morality in the new international order

If, however, it is utopian to ignore the element of power, it is an unreal kind of realism which ignores the element of morality in any world order. Just as within the state every government, though it needs power as a basis of its authority, also needs the moral basis of the consent of the governed, so an international order cannot be based on power alone, for the simple reason that mankind will in the long run always revolt against naked power. Any international order presupposes a substantial measure of general consent. We shall, indeed, condemn ourselves to disappointment if we exaggerate the role which morality is likely to play. The fatal dualism of politics will always keep considerations of morality entangled with considerations of power. We shall never arrive at a political order in which the grievances of the weak and the few receive the same prompt attention as the grievances of the strong and the many. Power goes far to create the morality convenient to itself, and coercion is a fruitful source of consent. But when all these reserves have been made, it remains true that a new international order and a new international harmony can be built up only on the basis of an ascendancy which is generally accepted as tolerant and unoppressive or, at any rate, as preferable to any practicable alternative. To create these conditions is the moral task of the ascendant Power or Powers. The most effective moral argument which could be used in favour of a British or American, rather than a German or Japanese, hegemony of the world was that Great Britain and the United States, profiting by a long tradition and by some hard lessons in the past, have on the whole learned more successfully than Germany and Japan the capital importance of this task. Belief in the desirability of seeking the consent of the governed by methods other than those of coercion has in fact played a larger part in the British and American than in the German or Japanese administration of subject territories. Belief in the uses of conciliation even in dealing with those against whom it would have been easy to use force has in the past played a larger part in British and American than in German and Japanese foreign policy. That any moral superiority which this may betoken is mainly the product of long and secure enjoyment of superior power does not alter the fact, though this consideration may well affect the appeal of the argument to Germans and Japanese and expose British and Americans to the charge of self-righteousness when they invoke it.

It is, however, useless to discuss these problems of power and morality in a nineteenth-century setting, as if some fortunate turn of the wheel could restore the old conditions and allow a reconstitution of the international order on something like the old lines. The real international crisis of the modern world is the final and irrevocable breakdown of the conditions which made the nineteenth-century order possible. The old order cannot be restored, and a drastic change of outlook is unavoidable. Those who seek international conciliation may study with advantage the conditions which have made the process of conciliation between social classes in some degree successful. Essential conditions of that process were that the reality of the conflict should be frankly recognized, and not dismissed as an illusion in the minds of wicked agitators; that the easy hypothesis of a natural harmony of interests, which a modicum of good will and common sense would suffice to maintain, should be consigned to oblivion; that what was morally desirable should not be identified with what was economically advantageous; and that economic interests should, if necessary, be sacrificed in order to resolve the conflict by the mitigation of inequalities. None of these conditions has yet been realized in the international community. Responsible British and American statesmen still commonly speak as if there were a natural harmony of interests between the nations of the world which requires only good will and common sense for its maintenance, and which is being wilfully disturbed by wicked dictators. British and American economists still commonly assume that what is economically good for Great Britain or the United States is economically good for other countries and therefore morally desirable. Few people are yet willing to recognize that the conflict between classes cannot be resolved without real sacrifices, involving in all probability a substantial reduction of consumption by privileged groups and in privileged countries. There may be other obstacles to the establishment of a new international order. But failure to recognize the fundamental character of the conflict, and the radical nature of the measures necessary to meet it, is certainly one of them.

Ultimately the best hope of progress towards international conciliation seems to lie along the path of economic reconstruction. Within the national community, necessity has carried us far towards the abandonment of economic advantage as the test of what is desirable. In nearly every country (and not least in the United States), large capital investments have been made in recent years, not for the economic purpose of earning profits, but for the social purpose of creating employment. For some time the prejudice of orthodox economists against this policy was strong enough to restrict it to half measures. In Soviet Russia, such prejudice was nonexistent from the outset. In the other totalitarian states, it rapidly disappeared. But elsewhere rearmament and war provided the first substantial cure for unemployment. The lesson will not be overlooked. A repetition of the crisis of 1930–33 will not be tolerated anywhere, for the simple reason that workers have learned that unemployment can be cured by a gigantic programme of economically unremunerative expenditure on armaments; and such expenditure would be equally effective from the standpoint of employment if it were devoted to some other economically unremunerative purpose such as the provision of free housing, free motor cars or free clothing. In the meanwhile we are moving rapidly everywhere towards the abolition or restriction of industrial profits. In the totalitarian countries this has now been virtually accomplished. In Great Britain, the assumption has long been made that to earn more than a limited rate of profit on the provision of essential public services is immoral. This assumption has now been extended to the armaments industry. Its extension to other industries is only a matter of time, and will be hastened by any crisis. The rearmament crisis of 1939, even if it had passed without war, would have produced everywhere changes in the social and industrial structure less revolutionary only than those produced by war itself. And the essence of this revolution is the abandonment of economic advantage as the test of policy. Employment has become more important than profit, social stability than increased consumption, equitable distribution than maximum production.

Internationally, this revolution complicates some problems and may help to solve others. So long as power wholly dominates international relations, the subordination of every other advantage to military necessity intensifies the crisis, and gives a foretaste of the totalitarian character of war itself. But once the issue of power is settled, and morality resumes its role, the situation is not without hope. Internationally as nationally, we cannot return to the pre-1939 world any more than we could return to the pre-war world in 1919. Frank acceptance of the subordination of economic advantage to social ends, and the recognition that what is economically good is not always morally good, must be extended from the national to the international sphere. The increasing elimination of the profit motive from the national economy should facilitate at any rate its partial elimination from foreign policy. After 1918, both the British and United States Governments granted to certain distressed countries 'relief credits', from which no economic return was ever seriously expected. Foreign loans for the purpose of stimulating production in export trades have been a familiar feature of postwar policy in many countries. Later extensions of this policy were dictated mainly by military considerations. But if the power crisis can be overcome, there can be no reason why it should not be extended for other purposes. The more we subsidize unproductive industries for political reasons, the more the provision of a rational employment supplants maximum profit as an aim of economic policy, the more we recognize the need of sacrificing economic advantage for social ends, the less difficult will it seem to realize that these social ends cannot be limited by a national frontier, and that British policy may have to take into account the welfare of Lille or Düsseldorf or Lodz as well as the welfare of Oldham or Jarrow. The broadening of our view of national policy should help to broaden our view of international policy; and as has been said in an earlier chapter,* it is by no means certain that a direct appeal to the motive of sacrifice would always fail.

This, too, is a utopia. But it stands more directly in the line of recent advance than visions of a world federation or blueprints of a more perfect League of Nations. Those elegant superstructures must wait until some progress has been made in digging the foundations.

*See p. 152.

Notes

1. P. Drucker, *The End of Economic Man*, p. 56.
2. Bernhardi, *Germany and the Next War* (Engl. transl.), p. 81.

3. *The Times*, 21 April 1939.
4. F. Naumann, *Central Europe* (Engl. transl.), pp. 4–5.
5. R. S. Baker, *Woodrow Wilson: Life and Letters*, v. p. 31.
6. *The Times*, 28 May 1935.
7. R. S. Baker, *Woodrow Wilson and World Settlement*, i. p. 18.
8. Hitler, *Mein Kampf*, p. 315.
9. *League of Nations: Eighteenth Assembly*, p. 49.

Index

Note: References in square brackets indicate that the subject appears in the endnote, not in the main text.

Printed in the United States
100487LV00003B/88-300/A